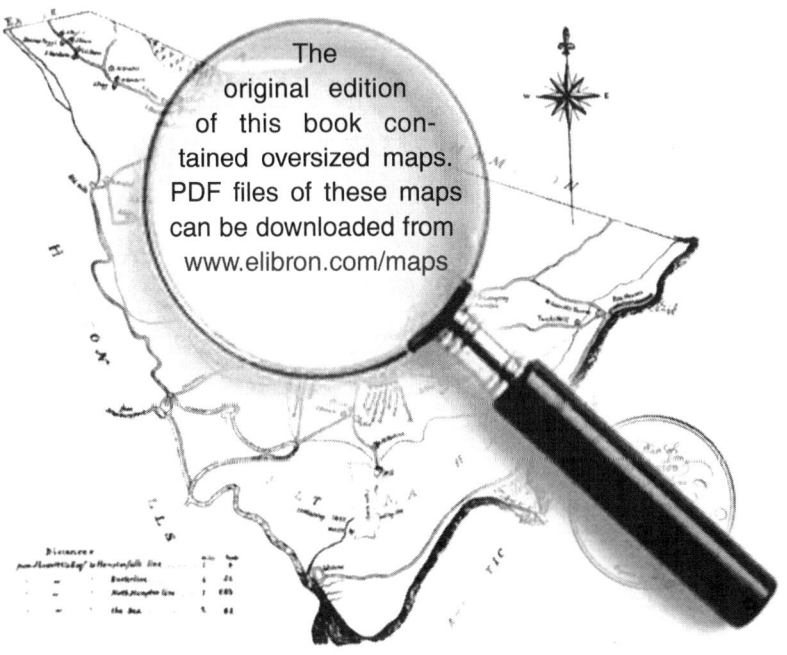

ROALD AMUNDSEN'S

"THE NORTH-WEST PASSAGE"

BEING

THE RECORD OF A VOYAGE OF EXPLORATION
OF THE SHIP "GJÖA"
1903-1907
BY ROALD AMUNDSEN

WITH A

SUPPLEMENT BY FIRST LIEUTENANT HANSEN
VICE-COMMANDER OF THE EXPEDITION

VOLUME I

Elibron Classics
www.elibron.com

THE "GJÖA" EXPEDITION,
1903–1907.

THE

NORTH WEST PASSAGE

BY

ROALD AMUNDSEN.

ROALD AMUNDSEN'S

"THE NORTH WEST PASSAGE"

BEING THE RECORD OF A
VOYAGE OF EXPLORATION OF THE
SHIP "GJÖA" 1903–1907 BY ROALD
AMUNDSEN WITH A. SUPPLEMENT
BY FIRST LIEUTENANT HANSEN
VICE-COMMANDER OF THE
EXPEDITION

WITH ABOUT ONE HUNDRED AND THIRTY-NINE
ILLUSTRATIONS AND THREE MAPS

VOL. I

London:

ARCHIBALD CONSTABLE AND COMPANY LIMITED

1908

Published November 23rd, 1907.

Dedicated

TO

H.E. Dr. FRIDTJOF NANSEN, G.C.V.O.,
NORWEGIAN MINISTER TO THE COURT OF ST. JAMES,

WITH THE DEEPEST GRATITUDE

FROM

ROALD AMUNDSEN.

PREFACE.

I TENDER my warmest and most heartfelt thanks to the small party of brave men who followed me through the North West Passage and risked their lives to ensure the success of my undertaking.

A loving thought will again and again travel back to the lonely grave looking out on the boundless ice-desert, and grateful memories will arise of him who laid down his young life on the field of action.

I tender my best thanks to Mr. Axel Steen, meteorologist, to His Excellency Professor Dr. G. von Neumayer, to Professor Ad. Schmidt, and to Professor Johannes Edler for the valuable assistance they rendered me with reference to the difficult magnetic problem connected with the Expedition.

The complicated financial accounts of the Expedition, as well as its voluminous correspondence, have been gratuitously attended to by Mr. Alex. Nansen, barrister, and my heartiest thanks are due to him for his valued services.

In conclusion, I beg to express my sincerest thanks to the Committee, who, on the initiative of Consul Axel Heiberg, and under his and Mr. Alex. Nansen's direction, have collected the necessary funds for meeting the liabilities of the Expedition.

I feel bound to add that Mr. Bernt Lie has very kindly revised my manuscript.

ROALD AMUNDSEN.

CONTENTS.

ILLUSTRATIONS.

Illustrations.

Illustrations.

THE NORTH WEST PASSAGE

BEING THE RECORD OF A VOYAGE OF
EXPLORATION OF THE SHIP "GJÖA"
1903 — 1907 BY ROALD AMUNDSEN
WITH A SUPPLEMENT BY FIRST
LIEUTENANT HANSEN
VICE-COMMANDER OF
THE EXPEDITION.

CHAPTER I.

INTRODUCTION.

FROM the days when the Phœnician sailors groped along the coasts of the Mediterranean, in the early dawn of civilisation, up to the present time, explorers have ever forged their way across unknown seas and through dark forests — sometimes slowly, and with centuries of intermission, at other times with giant strides, as when the discovery of America and the great voyages round the world dispersed clouds of ignorance and prejudice even in reference to the globe itself.

We all know that many explorers have been impelled by the desire for riches which they thought awaited them in unknown lands and seas ; in fact, it may be said of the majority of expeditions, that they would never have been undertaken had it not been for the

Chapter I.

stimulus of some purely material object or expectation. But the history of that branch of exploration, whose goal has been the eternal ice under the poles, shines forth with a bright and pure splendour of its own, not only with the lustre of the white snow-fields and strange celestial signs of the Arctic Region, but also with that of true and untainted idealism. If we except fishing or hunting expeditions pure and simple (to which, in fact, polar exploration owes a very great debt of gratitude), we may safely assume that even the most extravagant flight of imagination has never led anyone to penetrate the Arctic Regions in the hope of finding "gold, and green woods." It is in the service of science that these numerous and incessant assaults have been made upon what is perhaps the most formidable obstacle ever encountered by the inquisitive human spirit, that barrier of millennial, if not primæval ice which, in a wide and compact wall, enshrouds the mysteries of the North Pole. In spite of all the tragic calamities, the sad failures and bitter disappointments, these assaults have been repeatedly waged and as repeatedly renewed, and are still being renewed to this day. And this dogged perseverance, if it has not yet quite reduced the fortress, has at least forced its gates ajar and gained a glimpse into the far-distant mysteries lying beyond them.

A mighty breach was made in the ice rampart when Nordenskjöld achieved the North East Passage, and wrenched the Asiatic continent from its grasp. A generation earlier, Franklin, and the Franklin Expeditions, had proved that a strip of open sea bathed the whole coast of North America, and many are the other breaches made by skilled and hardy Polar explorers who have essayed to tear away the dark veil of mystery enshrouding the North; heavy have been the sacrifices made to achieve this end, and none more heroic than those made on the North West Passage.

Perhaps no tragedy of the Polar ice has so deeply stirred mankind as that of Franklin and his crew, stirred them not simply to sorrow, but also to stubborn resumption of the struggle. We knew there was a sea passage round Northern America, but we did not know whether this passage was practicable for ships, and no one had ever yet navigated it throughout. This unsolved question agitated above all the minds of those who, from their childhood, had been impressed by the profound tragedy of the Franklin Expedition. Just as the "Vega" had to navigate the entire passage to the East, so our knowledge as to this strip of open sea to the West must remain inadequate until this passage also had been traced from end to end by one ship's keel.

The little ship to whose lot this task fell was the "Gjöa." Little was it dreamt when she was being built, for a herring-boat, in the Rosendal ship-yard on the Hardanger, that she was to achieve this triumph, though it is hard to say what they do not dream of up there in the Fjords. Nor did any such dream ever enter the mind of her future skipper when the story of John Franklin first captivated his imagination as a

Chapter I.

boy of 8 or 9 years old. Yet the imagination of a boy offers a very wide scope.

May 30th, 1889, was a red-letter day in many a Scandinavian boy's life. Certainly it was in mine. It was the day when Fridtjof Nansen returned from his Greenland Expedition. The young Norwegian ski-runner came up the Christiania Fjord, on that bright sunny day, his erect form surrounded by the halo of universal admiration at the deed he had accomplished, the miracle, the impossible. That May day the Fjord celebrated its most beautiful spring revels ; the town was radiant with decorations ; the people held high holiday. That day I wandered with throbbing pulses amid the bunting and the cheers, and all my boyhood's dreams reawoke to tempestuous life. For the first time something in my secret thoughts whispered clearly and tremulously : " If *you* could make the North West Passage ! " Then came the year 1893. And Nansen sailed again. I felt I *must* go with him. But I was too young, and my mother bade me stay at home and go on with my lessons. And I stayed. My mother passed away, and for a time my affection for her memory struggled to keep me faithful to her wish, but at last it gave way. No bond could restrain my yearning to pursue the object of my old and only desire. I threw up my studies and decided to start the long training for the goal I had set before me, that of becoming an arctic explorer.

In 1894 I engaged as an ordinary seaman on board

Introduction.

the old "Magdalena," of Tönsberg, and went out seal-hunting in the Polar Sea. This was my first encounter with the ice, and I liked it. Time passed, my training progressed, and from 1897 to 1899 I took part, as mate, in the Belgian Antarctic Expedition under Adrien de Gerlache. It was during this voyage that my plan matured: I proposed to combine the dream of my boyhood as to the North West Passage with an aim, in itself of far greater scientific importance, *that of locating the present situation of the Magnetic North Pole.*

As soon as I got home I confided my project to my friend Aksel S. Steen, under-director at the Meteorological Institute. In fact, I did not know, myself, whether the object which I had set myself was of sufficient importance. He speedily convinced me that it was. With a card of introduction from Steen, I went to Hamburg to submit my project to the greatest authority of the day in terrestrial magnetism, Privy Councillor Professor Dr. G. von Neumayer, at that time Director of the "Deutsche Seewarte" (German Marine Observatory). As I unfolded my project, the amiable old gentleman's interest grew, until in the end he actually beamed with rapture, and for some time I received instruction at the "Deutsche Seewarte," under his personal direction.

At last the great day arrived when the project was to be submitted to Fridtjof Nansen. I think it is Mark Twain who tells of a man who was so small that he had to go twice through the door before he could be seen.

Chapter I.

But this man's insignificance was nothing compared to what I felt on the morning I stood in Nansen's villa at Liysaker and knocked at the door of his study. "Come in," said a voice from inside. And then I stood face to face with the man who for years had loomed before me as something almost superhuman : the man who had achieved exploits which stirred every fibre of my being. From that moment I date the actual realisation of the "Gjöa" Expedition : Nansen approved of my plans.

But, of course, that was not everything. For a Polar expedition I wanted money. And at that time I had very little. What I had at my disposal was barely sufficient to provide a vessel and scientific instruments, so the only plan was to go round to all who were likely to take an interest in the enterprise. This was "running the gauntlet" in a fashion I would not willingly repeat. However, I have many bright and pleasant memories from those days of men who encouraged me and gave me all the support they could. I have also other memories—of those who thought they were infinitely wiser than their fellow-creatures, and had a right to criticise and condemn whatever others undertook or proposed to undertake ; but let me put aside the dark memories and dwell only on the brighter. Professor Nansen was indefatigable in this matter as in all others. My three brothers helped me assiduously in my hard task. I first procured scientific instruments. Then it came to the *vessel*.

My choice fell upon the yacht "Gjöa," hailing from Tromsö. She was built, as I have already said, at

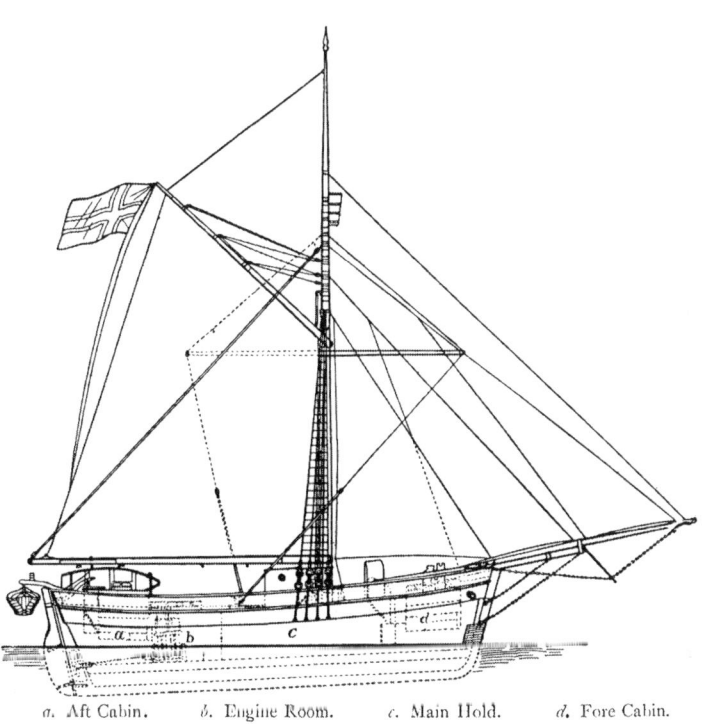

a. Aft Cabin.　　　*b.* Engine Room.　　　*c.* Main Hold.　　　*d.* Fore Cabin.

"GJÖA" (47 TONS R.).

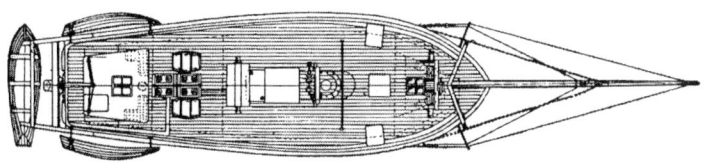

DECK OF THE "GJÖA."

Rosendal, on the Hardanger, in 1872. Her owner was Captain Asbjörn Sexe, of Haugesund. After she had been used for many years as a herring boat along the coast, she was sold in the eighties to Captain H. C. Johannesen, of Tromsö, and sailed the Polar Sea for some years. She was not spared there, but had ample opportunities of proving herself an uncommonly well-built boat. When I purchased the "Gjöa," in 1901, I had her fitted out for à summer voyage in the Polar Sea, so that I might test her and learn to handle her. I had never been on board a yacht, and was quite unaccustomed to handling so small a craft. The voyage turned out to my complete satisfaction ; the "Gjöa" behaved splendidly under all conditions. Of course, it was necessary to make a number of improvements before she undertook her proposed long voyage. Most of these were carried out at the Tromsö ship-yard, and I owe my deepest gratitude to the Works for the extreme con scientiousness with which everything was done.

In May, 1902, the "Gjöa" hoisted her flag and bade farewell to what for many years had been her home port. I put in to Trondhjem to have the necessary ironwork on board executed at the engineering works of Isidor Nielsen. Petroleum tanks were built to the shape of the boat. Our little motor—a 13 H.P. of the "Dan" type— which was connected to everything that could possibly be driven with its aid, was easy to work and practical in every part. The motor was the pet of every one on board. When it was not working we seemed to miss a

Chapter I.

good comrade. I may say that our successful negotiation of the North West Passage was very largely due to our excellent little engine.

In the spring of 1903, the "Gjöa" was berthed alongside the Fremnaes at Christiania to take in stores and provisions. The large, peculiarly built provision boxes were stowed and closely packed like children's bricks in a box. So neatly was this done that we found space on our little "Gjöa" for food and equipment, enough and to spare *for* 5 *years*. In the very important work of supplies I received invaluable help from Professor Sofus Torup. All our hermetically sealed goods, which were ready in October, 1902, were tested and examined by him. All our pemmican, both for men and dogs, was prepared by Sergeant Peder Ristvedt under Professor Torup's supervision, as was also the fish meal.

In May, 1903, the "Gjöa" lay ready for departure, and all who were to take part in the expedition assembled. Their names were :—

1. First Lieutenant Godfred Hansen, born in Copenhagen in 1876. He was second in command of the expedition. During his term of service in the Danish Navy he had made four voyages to Iceland and the Færöes and was warmly interested in Polar exploration. He was navigator, astronomer, geologist and photographer.

2. Anton Lund, first mate, born in Tromsö in 1864. He had served for many years as skipper and harpoonist on the Arctic Sea.

Anton Lund.   Helmer Hansen.
Godfred Hansen.   Roald Amundsen.   Peder Ristvedt.
Gustav Juel Wiik.   Adolf Henrik Lindström.

MEMBERS OF THE "GJÖA" EXPEDITION.

3. Peder Ristvedt, born in Sandsvær in 1873, took part, as assistant, on my trial trip with the " Gjöa " in 1901. He was our meteorologist and first engineer.

4. Helmer Hansen, second mate, born in Vesteraalen in 1870 and had served for many years in the Arctic Sea.

5. Gustav Juel Wiik, born in Horten in 1878. He had been trained at the Magnetic Observatory in Potsdam and was my assistant for magnetic observations. He was second engineer.

6. Adolf Henrik Lindström, born in Hammerfest in 1865, was cook to the expedition. He had served as cook in the " Fram's " second expedition.

As we had further financial troubles to face, it was not until June that all was settled and we were able to board our little craft and start on our voyage, and, following in the track of our predecessors, try to accomplish our task in the interest of human knowledge.

CHAPTER II.

Making for the Polar Sea.

The only thing that showed a visible sign of emotion at our departure was the sky. But this it did emphatically. When we cast off on the night of June 16th, 1903, the rain fell in torrents. Otherwise all was quiet that dark night, and only those who were nearest and dearest were gathered on the quay to say good-bye.

But in spite of the rain and darkness, and in spite of the last leave-taking, those on board the " Gjöa " were in high spirits. The last week's enforced idleness had tired us all. As to my personal feelings, I have neither the power nor the desire to give expression to them. The strain of the last days, getting everything in order, the anxiety lest something might yet prevent us getting away, and the desperate efforts to procure the money still wanting—all this had greatly affected me both in mind and body. But now it was all over, and no one can describe the untold relief we felt when the craft began to move.

Besides the seven participators in the expedition, my three brothers bore us company to see us clear of the Christiania fjord. All was hushed and quiet on board ; the navigation was attended to for the time by the steam

tug at our bows. The watch was entrusted to the helms-
man—and our six dogs. The dogs had formerly done
service in the second "Fram" expedition, which had
brought them home. Poor creatures! It would have
been better to have let them remain in ice and snow than
to drag them here, where they suffered sorely, especially
this spring, which was unusually warm. They were now
tied up along the rail and looked wretched in the rain—
the greatest infliction to an Arctic dog. To get here
they had made one voyage in the drenching rain, and
now they had to endure another to get back. But, at
any rate, back they were going—poor things—to their
home.

At 6 o'clock in the morning we put into Horten harbour
and took on board 4 cwt. of guncotton. Explosives *may*
be useful on a Polar expedition, and I should regard it
as a great mistake to start without them, even if, as
happened in our case, one has no occasion to use them.
At 11 o'clock A.M. we were off Faerder. The weather
had improved, and the rain had ceased. As we were
about to let go the tug rope it parted, and spared us the
trouble. The "Gjöa," under full sail, was making south
close to the wind, dipping her flag as a last salute to our
dear home. Long did we follow the tug with the telescope, and long did we wave our caps in answer to the
farewell signals which continued as long as the boat was
in sight. Now we were alone, and now we had to set to
work in earnest.

Deep-laden and heavy as she was, the "Gjöa" did

Chapter II.

not show much speed. As all had been made ship-shape beforehand, we were able to begin our regular duties at once. The watches were kept and relieved in turn. It was glorious. No anxiety, no troublesome creditors, no tedious people with foolish prophecies or sneers; only we seven joyous and contented men on board, where we wished to be, and with bright hope and confident trust in the future we were going out to meet. The world which had been oppressive and gloomy so long seemed to be again full of spirit and delight.

The last we saw of land was the Lister light. In the North Sea we encountered a couple of gales none too agreeable to those who were not sea-proof. The dogs were now let loose and jumped about as they liked. On days when we had heavy seas and the " Gjöa " rolled— and she can roll—they would go about studying the faces of the various members of the expedition. The rations they had—one dry cod and two pints of water— were not sufficient to satisfy their appetites, and they resorted to every possible expedient to get an extra meal. They were old acquaintances and agreed fairly well, at any rate, as far as concerned the male element. With the two ladies—Kari and Silla—matters were different. Kari was the elder and exacted absolute obedience, which Silla, who was also a grown-up lady found it difficult to put up with, so it often happened that they tore each other's hair. Ola, who was acknowledged to be the boss, did his best to prevent such battles. It was a rare sight to see old Ola—intelligent to an exceptional degree

—jumping about with the other two, one on either side, trying to prevent them fighting.

Our daily routine was soon working smoothly, and everyone gave the impression of being eminently fitted for his post. We constituted a little republic on board the "Gjöa." We had no strict laws. I know myself how irksome this strict discipline is. Good work can be done without the fear of the law. As the result of my own experience I had determined to apply the system of freedom on board as far as possible, to let everyone feel that he was independent in his own sphere. This creates among sensible people a voluntary spirit of discipline, of far greater value than enforced rule. Everyone thus obtains the consciousness that he is a man who is depended upon as a thinking being and not as a wound-up machine. The zeal for work is doubled, and so is the work itself. I can recommend to all the system adopted on board the "Gjöa." My comrades also seemed to value it, and the voyage of the "Gjöa" was far more like a holiday trip of comrades than the prelude to a serious struggle lasting for years.

On June 25th we passed between Fair Isle and the Orkneys out into the Atlantic. The many who had predicted our destruction ere we reached here, should have seen us now, under full sail and a fresh breeze from the south-east, making at high speed for the west. The "Gjöa" vied with the sea-gulls in dancing on the crests of the billows. Yet it was remarkable how devoid of life was the sea. We saw neither bird nor fish, to say nothing of sailing craft. We had only seen one full-rigged vessel since we took bearings off Lister.

The engine came in useful several times. When the wind dropped, so that we were only making under two knots, I decided to put it in action. It was, however, important to economise the petroleum as much as possible, as we could not tell how long the voyage might last.

Everything was now in good trim and running smoothly. The day was divided into four six-hour watches, three men to each watch. The service was shared equally among us all ; when the motor was operating, the engineers were mostly in the engine-room. Yet they were always ready to help us deck hands at a pinch. The old feud between deck and engine-room hands was unknown on board the " Gjöa." We all worked with a common object, and willingly and cheerfully took part in everything. Thus, as a rule, two men remained on deck, and we each took our turn at the helm.

At the end of July sickness began to appear among the dogs. It seemed to affect them mentally at first. They went about in a stupid state. They neither saw nor heard. They had little or no relish for their food. After a couple of days their hindquarters became paralysed, so that it was only with great difficulty they could drag themselves along. Finally they had convulsions, and we were glad to end their lives with a bullet. In this way we lost two fine beasts, Karl and Josef, to the great satisfaction, be it said, of Silla, who now remained cock of the walk.

At first we proceeded, as much as possible, by great circle sailing. The weather had been favourable and our progress unexceptionable. On July 5th we caught a light breeze from the S.S.E. We were running with peaked sails, and cut through the water at a speed of 10 knots. The main boom was well eased off and the stopper made fast. It was drizzling when I went to my bunk in the evening. At 1 A.M. the wind jumped round to the east, with the result that the main sail swung over. The boom stopper broke and the boom swung round with terrible force. This might have had serious consequences, but luckily, simultaneously with the stopper, the belaying-pin to which the peak-halliard was fastened also broke, with the result that the peak came down and deadened the blow which otherwise might have cost us our boom. It was a comparatively cheap experience. We were more careful at night after that.

Our four surviving dogs meanwhile began to be manifestly bored. In the beginning they could study wind and weather and thus kill time ; but now, meteorological variations failed to interest them, and their thoughts sought new fields. Idleness is the root of all evil, it is said, and this applies just as well to beasts as to men. Lurven and Bismark, who so far had been quite devoted and loyal to Ola, now began to make objections and dispute his authority. In other words, Lurven, who was born wicked, stirred Bismark up. The latter was a big, fine dog about two years old, with the most splendid set of teeth I have ever seen. Age had left their mark on Ola's teeth, and they were now the worse for wear. As their ancient leader, he still inspired respect, and the others thought twice before attacking him. Lurven, however, played his part very cleverly. He would start at full speed towards Ola. Bismark, who perceived that an assault was imminent, instantly joined his companion to assist in the undertaking. When right up to Ola, Lurven would pull up, consequently Bismark, who was unprepared for this stratagem, ran straight into the enemy's mouth. As a rule, he would then get a good drubbing from the more experienced Ola. Lurven was the most mischievous dog I ever knew. I can see him now with head on one side, his little eyes blinking and tail cocked sideways, gliding along the deck meditating some new prank. He often got a licking from us for his nasty tricks, and accordingly deferred his operations to a time when he was less observed. If, for instance, we were busy with the sails, we might be quite sure of a fight. In the pitch darkness of the night, when he had brought the others together, he often seized the opportunity to fall upon Ola in the rear, and then the old dog was overmatched. Poor old Ola, he was often roughly mauled in these nightly contests. On these occasions Silla would jump round the combatants making the most deafening noise and, by way of variety, snapping at their legs.

It rained continually and steadily, and we collected

water in all our vessels for washing and to water the dogs. But, as a rule, we washed in salt water and did not make such a point of being absolutely white.

AN ICEBERG IN BAFFIN'S BAY.

We now kept a sharp look out for ice, and on July 9th we set eyes on two narrow strips, which lay undulating in the sea, showing us that we might now shortly expect the main mass of the ice. And so it was. Not long after we saw pack ice ahead in great dense masses. In its wake came the fog, ice's faithful attendant, and it kept

Chapter II.

us company during the greater part of our navigation in Arctic waters.

On July 11th, at 2.30 P.M., we sighted land, a little to the west of Cape Farewell. The high, craggy mountain landscape stood out splendidly. The ice seemed to lie quite close up to the shore. Following the advice of the Scotch whale-fishers, Milne and Adams, I stood well off the shore, to avoid getting into the ice. On the 13th we met the first icebergs, two solitary majestic masses. Those of us who had not seen such things before were naturally much interested, and the telescopes were in great demand.

At the sight of the ice the hunter's blood began to stir in most of us. A look-out was kept with the telescopes for possible booty, and bear stories were a constant subject of conversation. Of course, Bruin stood highest in the list of anticipated adventures; but we should also have welcomed a crested seal, the big, handsome seal which is found on the ice along the Greenland coast. A couple of the mightiest hunters whispered softly of the possibility of killing a whale.

At last, on July 15th, our hunters tasted their first blood. We made that day a short detour into the ice, and shot four big crested seals. The fresh flesh tasted excellent. Lindström talked volubly of rolled meat, brawn, and sausages till our mouths watered. He bragged of his culinary exploits as *chef* on the " Fram." Unfortunately, his tale concerned only the past, and we expected and hoped to see his deeds in the present.

But, so far, in vain. Well, honour to the "Gjöa's" cook; he has pickled many a good piece of beef for us, anyhow.

It was, however, not only we men who relished fresh meat. The dogs exhibited the most manifest proofs that they were equally fond of good fare. They stuffed till they were as tight as drums; especially did Lurven distinguish himself. He was now a sorry sight: smeared all over with fat and blood. And the scavengers, who had hard work at the best of times during every watch, more than earned their wages in clearing up after such a feast. Every sailor is familiar with *that* result of having a dog on board. So just imagine four dogs at once, quite devoid of any polite training.

The next day we were again in the ice, and shot seven more seals. Harpoons and knives had very hard wear in that seal season. Our inventive engineer, however, had hit upon the idea of driving the grindstone by means of the sounding apparatus, which was worked by the shaft, so he managed to do all the grinding without any assistance.

Lindström thought seal liver one of the greatest delicacies in existence, and he treated us to it morning and night. It must be owned it does not taste badly at all. When we came near " Lille Hellefiskbank " (Little Halibut Bank), the engineer, who was as enthusiastic a fisherman as he was a hunter, got his fishing tackle in order, and installed himself in the stern, whence he carried on his fishing in grand style. He was himself very sanguine and was backed up by the cook, but we were very sceptical. Great was his triumph when one morning he actually caught a little halibut, which tasted beautifully.

On July 20th we made the acquaintance of the "Sugarloaf." The coast still generally maintained its character with high craggy summits. However, it was more lively here, with shoals of whales every now and then. The weather was also better shorewards than out at sea ; it kept clear with a light breeze from the south.

AIRING MOULDY BREAD.

The temperature of the water was quite 39° Fahr. Strangely enough we saw no more ice, although one would think that the steady north wind we had had against us all the time would have brought a quantity of ice southwards. Could it be, perhaps, that there was no more ice ?

We now noted for the first time that the compass was not to be relied upon. This is, however, a well-known phenomenon here on the west coast of Greenland. Further out at sea it is trustworthy enough. The large amount of iron in the mountains is probably the cause.

July 24th was a beautiful day, dead calm and dazzlingly bright. It was the first real summer day we had had since our departure. We seized the opportunity to bring up into the air all the bread which we had brought fresh from home and had spread out below in the hold. Much of it was spoilt ; but we cut the mould away and aired what was left whenever we had a chance.

" A sail ahead ! " is the sudden cry ; and there was life aboard. All telescopes—and we had a lot aboard the " Gjöa "—came out.

" A full-rigged ship ! " someone exclaimed.

Well, we shall be content with a schooner, thought I.

" I see clearly, it is a brig."

" It is probably one of the Royal Danish Greenland Trading Company's ships, on her way home."

" There's another ! " shouts one with a telescope to his eye.

Well, it is beginning suddenly to be populous here in the ice waste. And we stroll up and down the deck and mutter in very high spirits what a surprise we shall be for the new arrivals. We must admit that we even smartened up the deck a little. We might have a visit.

Then there was a telescope that was shut up in a decisive manner, followed by a burst of laughter.

" Well ? "

" Gentlemen," says Lieutenant Hansen, " it is an iceberg ! "

Chapter II.

Then follow indignant protests, spying through the glasses and discussion, while we continued to approach the object of dispute. The heat of discussion disappears, the full-rigged ship is abandoned, and after it the brig. The schooner still had one adherent even when we had advanced near enough to see clearly before us a great gathering of icebergs ; they seemed to be aground on the "Store Hellefiskbank" (Big Halibut Bank).

Later in the forenoon we sighted Disco Isle, high and flat-topped ; easily recognisable at a long distance. But it takes a long time to get to it. At 8 o'clock in the evening we were still 30 miles off, and only at half-past ten the next forenoon did we get in near the shore. A barrier of grounded icebergs seemed to block the entrance to Godhavn. But soon Nielsen, the governor of the colony, came out to us in a boat to bid us welcome and pilot us in. We met a violent squall and had to tack, as the motor could not manage it alone. At 1 o'clock we anchored.

Godhavn lies on a small, low island, separated from Disco by a very narrow sound. In 1903 the town numbered 108 souls. It is the residence of the Inspector of North Greenland. It is beautifully situated with the lofty and mighty Disco to the north and the sea to the south and west, from time to time filled with heavy icebergs.

We at once paid visits to the authorities of the place, the Inspector and the Governor of the colony. In the previous year I had been in correspondence with Inspector

GODHAVN, WITH DISCO ISLAND IN THE BACKGROUND.

Daugaard-Jensen, who had promised to get me ten dogs, with complete outfits. He received us with great amiability, and was able to inform us that all had arrived in good condition—sledges, kayaks, ski, 20 barrels of petroleum, etc. The Royal Danish Greenland Trading Company had shown us the great favour of loading these things for us on one of their ships. I owe Director Rydberg and Office Manager Krenchel my best thanks for the exceptional treatment accorded to the "Gjöa" Expedition by the Company.

Nielsen was untiring in the help he afforded us in every respect. We divided at once into two parties, one to carry out the necessary observations, while the other took charge of all the work on board. Lieutenant Hansen superintended the astronomical and Wiik the magnetic observations. Lund and Hansen were to get all on board, and, in general, to get the ship ready to continue our voyage. Ristvedt went backwards and forwards, and had his hands full. At one time he had to take readings of the chronometer for the astronomer, at another for the magnetician; now he was in the hold looking after the water-tanks, now in the engine-room drawing petroleum. It was a busy time. But how we worked! All seemed inspired with the same zeal to get the work done well and quickly, so that we might get away as soon as possible and lose no time or opportunity for progress.

Lindström knew how to make matters go smoothly— in his own way. He was busy everywhere, bought from

Chapter II.

and bartered with the Eskimo, now a salted salmon, now a fresh one, now an eider duck, now a loon. So at that time the bill of fare was well varied. Lindström's coin was baker Hansen's mouldy spiced cakes from Christiania. If this coin had not the right ring or was not quite up to the standard, it was, at any rate, both round and current. When an Eskimo came to sell, Lindström was fetched on deck. The negotiations were carried on in Eskimo and good Nordland Norse. The retorts which fell from both sides were both long and smart, but those from the Eskimo grew more obsequious and timid as compared with the fatherly condescension of Lindström, who did not seem to want or wish for anything in the world. We who knew that our dear cook had not the faintest notion of Eskimo, gathered in couples ready to burst with laughter. When the discussion had lasted some time, Lindström would suddenly show a bright gleam of intelligence and disappear into the hold. Big-looking and benign he would come back again, with a mouldy spiced cake under each arm. The Eskimo regarded Lindström with an expression of the liveliest astonishment, as he asked for tobacco in exchange for his salmon. Any attempt to make Lindström understand his error is met with a thorough, condescending, shoulder-shrugging non-comprehension. Lindström takes the salmon, the man takes the cakes, and the transaction is finished. The epilogue is, perhaps, after all the best of it : to hear Lindström relate that he, of course, understood every word uttered by the Eskimo,

ESKIMO BEAUTIES FROM GODHAVN.

"but as he asked for three cakes, I pretended I did not understand, and gave him two." I had my dark suspicion that the Eskimo had taken cakes home to his own people more than once, and, undoubtedly with better reason, boasted to them that "he had pretended he did not understand."

Our stay in Godhavn was altogether a pleasant one. Our greatest plague was the gnats that worried us to such a degree in our work that from time to time we had to fly below into the cabin to get a little peace. It was seldom our tormentors followed us there.

On July 31st we were ready. The various observations had been taken, and all our outfit was aboard. We had no time to spare, and had to hurry, so we took leave of the amiable people in Godhavn and weighed anchor. The Inspector, the Governor, and his assistant accompanied us out of the Sound. The public buildings displayed bunting, and a salute thundered from the battery on the hill. Among the belt of skerries around the shore we said a last farewell to our friends, again saluted the hospitable Danish flag, and we were again left to ourselves. As soon as we were out at sea we met our old friend the nor'wester, and had to tack. Parry Skerry was marked wrongly on our chart, and we barely escaped running on to it. Luckily we saw the breakers on it, and turned away in time. It is quite low, and has a striking resemblance to a whale's back.

During the stay at Godhavn I had served out some of our thick underclothing—Iceland jerseys and " Nansen "

Chapter II.

clothing—to each of our company, so that we might be
ready to meet the ice. Most of us had also procured
sealskin clothing by barter.

On August 6th we were abreast of Upernivik,
12 miles off. Here hundreds of icebergs had collected.
They looked larger and more compact than those we
met to the south. Of drift ice we had not yet seen a
trace, and we began to entertain the hope of slipping
unhindered across Melville Bay. The day after we
passed Itivdliarsuk in latitude 73° 30′ N., the northernmost spot inhabited by civilised men. On August 8th
we were off Holms Island, and were about to begin the
voyage across Melville Bay. This is the most dreaded
stretch in that part of the Arctic Ocean. Many are the
vessels that have made their last voyage here. It is,
however, earlier in the year that the conditions are
especially dangerous. In June and July, when the ice
breaks up and the whale fishers go north—it is, of course,
important to be first in the field—they often have a
severe struggle with the ice. The outer part of the ice
in the bay breaks first, and the inner part remains quite
whole. This is what is called shore ice or fast ice.
Along its edge the whalers seek a passage, and the wise
ones among them do not quit it till they are out in
the open water on the north side of the bay. On the
borders of the shore ice there are often formed natural
havens where vessels can run in for shelter when the
drift ice sets in. If there be no natural dock, most of
the whalers have sufficient crews to cut their way into

the ice in a comparatively short time. It is the Scotch who reign supreme in these waters ; and there is no doubt that these Scotch whalers, under dangerous and difficult conditions, have become some of the doughtiest Arctic seamen of our time.

At Holms Island we set course for Cape York. The state of things seemed very favourable. No fast ice was to be seen, and as far as the eye reached, Melville Bay was filled up with icebergs and "knot ice," *i.e.*, fragments of icebergs. At 3 o'clock in the afternoon we passed the well-known landmark called the "Devil's Thumb," a mountain peak with such a striking resemblance to an old, rugged, up-turned thumb that we all burst out laughing at the sight of it.

We now set all sail and put the motor at full speed. It was important to get as quickly as possible across the bay and to spare nothing. But, alas, our course towards Cape York was not of long duration. Next morning we were stopped by thick pack ice. During the night a quarter of an inch of new ice had formed, so we were obliged to accept our ill luck and turn southwards, like so many before us. However, we first turned aside into the ice to get a nearer look at it. The even surfaces and sharp edges indicated that it was newly broken-up shore ice ; probably we had kept too near the shore. We now followed this ice towards the south-west. A tongue of ice stretched out towards the south-west ahead of us. The atmosphere was gloomy above it, and indicated open water. However, behind that tongue of ice there was another, on the opposite side of a large bay full of drift ice. We tried to force our way into that bay, but the ice soon got more tightly packed, and pressed us out again. Further out the ice was considerably heavier, and it looked as if we were just on the boundary between the newly broken-up shore ice and the drift ice. I, therefore, resolved to keep on the move to and fro' here, where probably any change in the conditions would at once manifest itself. And I was right. At midnight the ice slackened and let us slip in without any particular trouble. At the same time a fog set in, pitch dark. Those who have not seen the ice fog of the Arctic Ocean do not know what fog is. London fog is nothing to it. We could not see the ship's length. But we held our course with the aid of the compass, and the ice politely made room for us. Thus we went forward through the damp fog; but if anyone asked me about the ice conditions in that part of Melville Bay, I could not give any information whatever. The monotony was broken now and again by a seal which at once paid forfeit with its life; we revelled in fresh seal's flesh. We had not seen a bird all the time, but we noticed what seemed to be quite large coveys of little auks that flew in thousands right by the ship. One great advantage in drift ice is that it affords an abundant supply of water. On almost every floe there is a pool of the most beautiful drinking water, and we even allowed ourselves the luxury of washing and bathing in fresh water.

Making for the Polar Sea.

On August 13th, at half-past two in the morning, I stood at the helm, after relieving the watch at two, shivering with cold. Perhaps, as an Arctic traveller, I ought not to admit this, but anyhow, I did feel perishing with cold. My two companions in the watch strolled about the deck and tried to keep warm as best they could. The fog settled down and drenched everything it came in contact with; it was sheer misery in the early morning. The watch below were now enjoying steaming hot coffee which they well deserved after a spell of six hours' duty. Suddenly a gleam of light broke through the fog, and, as if by enchantment, there opened up before me a wide view out into the bright daylight; right in front of us, and seemingly quite near, the wild, rugged landscape of Cape York appeared suddenly like a scene from fairyland. We all three cried out simultaneously with surprise and delight. The watch below came rushing up from their coffee, and soon all hands stood in rapt and silent contemplation of the scene. The morning was so dazzlingly, supernaturally clear that we imagined we could make Cape York in a couple of hours' sail. And yet it was 40 miles off. To the east the whole interior of Melville Bay lay before us. Right inside, in the farthest background, we could see several mountain tops. An impenetrable mass of ice filled the bay; mighty icebergs rose here and there from out of the mass of ice. When at last we looked back, we saw the fog out of which we had suddenly slipped, lying thick like a wall behind us. Such a sight is one of those wonders only to be seen in the never-to-be-forgotten realms of ice, and make us long to return and feel again their enchantment, in spite of all toil and privations.

The ice conditions in the direction of our course looked promising. There certainly lay some ice to the windward, but we did not pay much attention to it. However, on the same day, at noon the ice closed in, so that there only remained an open channel towards the north. We were then 25 miles from Cape York. However, the ice gradually slackened ahead of us, smoothing the way, and at 5 o'clock in the afternoon we reached the edge of the shore ice under Cape York. Then for a time we headed for Cape Dudley Digges. As fog now set in again, we made fast to the ice, to wait till it lifted. Two of our hunters availed themselves of the opportunity to go out in a boat to shoot small auk. After a couple of hours they came back, with birds enough for a dinner. They tasted like the most delicious fieldfare, and it is wonderful what a gourmand one becomes on a voyage in the Arctic Ocean.

When the watch was being relieved next morning, it cleared up. Our near surroundings were rather thickly packed with ice. However, a mile south of us there was a large broad opening in the ice going westward, and though unwilling to go back, I thought it the most advisable course. After a lot of toil we got out into the opening. It widened out more and more to the westward, and there was no doubt about its leading out into open water. And, indeed, at half-past three we were in a sea free from ice.

Melville Bay had been conquered. We had every reason to be pleased ; that portion of the sea had always appeared to me as the most difficult to get through, with such a small ship as ours, in the whole North West Passage. And now we had navigated it without mishap.

At four o'clock on August 15th we reached Dalrymple Rock, where the Scotch whalers, Captains Milne and Adams, had deposited considerable stores for us. Dalrymple Rock is easily recognisable from the descriptions. It rises right up from the sea in a cone. When approaching it, as we did, from the east side of Wolstenholme Island, one first catches sight of a little island to the north of it. This is Eider Duck Island. Here and on Dalrymple Rock the Eskimo gather large quantities of eggs every year.

"Two kayaks ahead!" suddenly bawled the look-out a-top. In a trice all hands were on deck. I stopped the engine and the kayaks came close alongside. We were very anxious to make the acquaintance of the North Green-land Eskimo, of whom many strange things are reported. They were two really good-looking men. Their costume seemed to us somewhat strange at first sight, and there was no end of laughter when one of them stooped to pick up a knife he had dropped, and in doing so showed that a stitch in time might have saved nine. A pretty bow, indeed ! They were extremely lively, jabbered both to-

gether, threw their arms about, and gesticulated. There
was evidently something particular they wanted to tell us.
But we, of course, could not understand a syllable. So

THE DANISH LITERARY GREENLAND EXPEDITION.

one of them put on a broad grin, and said : " Mylius ! "
And then it dawned upon us what it was all about. The
Danish so-called Literary Expedition to Greenland, under

Mylius Erichsen, must be in the neighbourhood. According to what we had heard of it, we thought it must be with the Cape York Eskimo. Scarcely had the name been uttered, when there burst forth a rattle of fire-arms, as if a regular battle were raging behind a piled mass of ice, and six kayaks issued from among it like flashes of lightning. One of the kayaks was decked with a little Norse flag, and another with the Danish flag. It was in truth a pleasant surprise.

We soon had on board the leader of the expedition, Mylius Erichsen, and one of its members, Knut Rasmussen, together with four Eskimo. They were made welcome, and question and answer tumbled over each other pell-mell, in joyous confusion; this lasted some time until both sides quieted down sufficiently to exchange some sort of reasonable information. Our first anxiety was the depôt, and to our relief we learnt that it was in capital order. At 7 o'clock in the evening we reached Dalrymple Rock. There is no harbour in the island, so we had to lie unprotected. However, I rowed at once ashore with Lund to inspect the depôt, and determine how we must set about getting it aboard. Mylius Erichsen handed me a letter from Milne and Adams, in which they wished us all luck on our voyage. I cannot sufficiently thank the two gentlemen for the readiness with which they undertook the tedious work, and the care with which it was carried out. The depôt lay among big stones on a slope of a ridge, and was surrounded on all sides with barbed wire. At the end of the ridge a footing of old ice projected out into the sea, and formed the most beautiful natural quay. We therefore decided to rig up our derrick on the quay as a crane, and with its aid to put the cases right into the boat after we had brought them up in sledges. So as not to make the transport by boat too long I took the " Gjöa " as close into the shore as possible and anchored there. I admit that that was imprudent on an open coast, but it was of importance for us to get away quickly. We sent a boat to the shore to fetch Count Moltke, the third member of the Expedition, who was lying ill.

We snatched a hasty supper, and at 10 o'clock we set to work. Lieutenant Hansen remained on board to superintend. I myself undertook the work on shore with the kind assistance of our Danish guests and some Eskimo. Hansen was to ship the cases, and Lund was to heave them on board. The whole depôt, 105 cases, had to be taken as deck cargo. Meanwhile the motor was cleaned and polished by Ristvedt and Wiik.

At 2 o'clock in the morning we took a little rest and a cup of coffee, which we were greatly in need of. The cases weighed about $2\frac{1}{2}$ cwt. each on an average, so it was no child's play. At half-past two we had the satisfaction of seeing Count Moltke join our company. After the coffee we set to work again. I now had four Eskimo to help me. It has often been said that the Eskimo are lazy, unwilling, and possessed of all other bad qualities under the sun. Certainly this was not true as

to these four helpers of mine. They handled our cases,
many of which weighed nearly 4 cwt., with an ease

NORTH GREENLAND ESKIMO.

and skill which would be hard to match. Instead of the
oaths and execrations, which among "civilised" work-

43

men are wont to accompany such work, these children of Nature carried out their task with song and merriment.

At 8 in the morning the last cases, together with six barrels of petroleum, were brought down to the quay, and I expected that we should be quite finished by 9 o'clock. But, alas! I had miscalculated. All at once there sprung up a sea breeze, which compelled me to go aboard in a hurry. The anchor was weighed and the foresails set; stopping to get the mainsail up was out of the question. The squall became very strong, but luckily the wind jumped round. so that it filled our sails. Now we went ahead at a good speed, and it was high time, as our distance from the shore was to be measured by inches. We sailed round the island and anchored on the lee, on the other side. But now we had the fatiguing work of transporting the eleven cases and six barrels of petroleum left on the quay over to the opposite side of the island. I dreaded proposing this to the Eskimo, but they only joked and laughed and set to work as if they had only just begun. It was 7 o'clock in the evening before we had finished.

On our arrival at the island the dogs were let loose so as not to be in the way during the work. They employed their time well. The old "Fram" dogs and the new ones from Godhavn seized the opportunity to settle up, in a battle royal, all the quarrels on board they had nursed up to date. Many of them bore dreadful marks of the battle when they were brought aboard again. One of our new dogs declined to come on board and we had to leave him behind. The Eskimo would be sure to catch him when he got hungry. Mylius Erichsen gave me four spendid dogs, two full grown and two puppies, a couple of months old. These two puppies became unusually fine dogs. We called them "Mylius" and "Gjöa," and the latter was undoubtedly our very best dog. At 11 o'clock in the evening we reached Saunder Island where the Literary Expedition was staying, and although it was very hard to say good-bye to them there, so soon after meeting them, we were compelled to do so.

We were now heavily laden. Our stock of petroleum on leaving Dalrymple was 4,245 gallons. The deck was down to the waterline, and the cases were piled almost up to the main boom. The dogs got on the tops of the cases and waylaid each other. We had rare trouble to keep the two hostile parties from fighting.

At 2.30 in the morning of August 17th we continued our voyage. It was a magnificent morning. Glacier upon glacier shone far towards the north, until the land ended at Cape Parry. At the sight of the glacier, where our brave countryman, Eivind Astrup, halted with Peary to begin his journey over the inland ice, I found it hard to take my eyes and thoughts away from it ; but I had to do so and fix my attention on my own affairs. Ahead of us stretched a wall of heavy, newly formed icebergs which we had to keep clear of.

Greenland now began to dwindle away and we stood

Chapter II.

well in towards Cape Horsburgh, the northern entrance to Lancaster Sound. Later on in the day we passed the Carey Islands at a distance of 15 miles. Luckily the weather kept calm and clear. Loaded down as the " Gjöa " was, we were not fit to battle with a storm. It was a stiff job to get round Cape Horsburgh. The wind had quite dropped, and a heavy swell from the south, meeting the current out of the sound, caused a very nasty sea. And the " Gjöa " was not a flier when travelling under engine power. At last on August 20th, at half-past four in the morning, we rounded the cape and got into Lancaster Sound. As I had determined to go to Beechey Island in order to undertake a series of magnetic observations there, we kept close in under the northern shore. With the exception of a few icebergs and a little loose ice, extending from the shore, the fairway was very nearly free from ice. Fog followed us all the way to Cape Warrender. Here it lifted, and in the fine clear weather we could observe the land. It presents a very different aspect from that of the wild and rugged look of Greenland's mountains. The most prevalent form is the plateau, but it is very often broken by dome-shaped heights ; it is barren, but yet not unpleasing. The clear weather did not last long. By the following morning the fog hung over us again. The compass was now rather unreliable, and this, combined with the fog, must serve as our excuse for having made a mistake here ; we did so twice. But I console myself with the knowledge that the same fate may befall those who come after us.   After rather hard tacking we reached Beechey Island on August 22nd, at 9 o'clock in the evening, and anchored in Erebus Bay.

By the time the anchor was down and the vessel hove to, most of us had turned in to enjoy a night of unbroken sleep.   It was about 10 o'clock when twilight came on. I was sitting on one of the chain lockers looking towards the land with a deep, solemn feeling that I was on holy ground, Franklin's last safe winter harbour.   My thought wandered back—far back.   I pictured to myself the splendidly equipped Franklin Expedition heading into the harbour, and anchoring there.   The " Erebus " and " Terror " in all their splendour ; the English colours flying at the masthead and the two fine vessels full of bustle—officers in dazzling uniforms, boatswains with their pipes, blue-clad sailors ; two proud representatives of the world's first seafaring nation up here in the unknown ice waste !

A boat is lowered from the commander's ship ; Sir John is going to land.   The men pull hard, proud of having the commander in the boat.   His clever face, full of character, beams with gentleness ; he has a word for everyone, and is therefore loved by his men.   They feel an unbounded confidence in the experienced old leader in Polar lands.   How they listen attentively to every word that passes between their chief and the two officers between whom he is seated.   The conversation concerns the unfavourable state of the ice and the possibility of wintering at Beechey.   Sir John finds it hard to reconcile

Chapter II.

himself to such an idea. But from old experience he knows that in these regions one is often compelled to act very much against one's will.

Certainly these brave men had succeeded in discovering much new land, but only to see their expectations of the accomplishment of the North West Passage that way brought to nought by impenetrable masses of ice. The winter of 1845—1846 was passed here on this spot. The dark outlines of crosses marking graves inland are silent witnesses before my eyes as I sit here. The spectre of scurvy showed itself for the first time, and claimed, if not many, yet several lives.

At the breaking up of the ice in 1846, the " Erebus " and " Terror " again stood out to sea. Once more resounds the merry song of the sailors, and the vessels pass out between Cape Riley and Beechey. Once more waves England's proud flag ; it is the farewell of the Franklin Expedition From this point it passed into darkness—and death.

The great explorer, Dr. John Rae, was the first to bring news as to the region where the Franklin Expedition was lost. But the honour of having brought the first certain intelligence as to the fate of the whole expedition belongs to Admiral Sir Leopold M'Clintock. So many books of travel contain accounts of the tragedy that I will not repeat the story. Franklin and all his men laid down their lives in the fight for the North West Passage. Let us raise a monument to them, more enduring than stone : the recognition that they were the first discoverers of the Passage.

August 23rd brought fog. Wiik and I at once began the magnetic observations. They were watched this time with great excitement and interest by all on board. Indeed, our route to the Magnetic Pole depended on their result. It must not be denied that many hoped that the compass would point westward, towards the musk-oxen on Melville Island and Prince Patrick's Land. The dipping needle was released, and its movements were followed with breathless suspense. It oscillated long, and at last came to rest in a south-westerly direction. Although at times I also had thought with pleasure of the hunting fields in the north-west, I felt very satisfied now that the point was decided. My original plan could be continued, and my comrades were inspired with the same feeling. We were always agreed that the best route for the North West Passage must be the very one the magnetic needle now indicated.

Wiik was a steady worker. A more conscientious and careful assistant I could not have had. Lieutenant Hansen could not make use of his astronomical instruments. The sun would not come out, and we had to satisfy ourselves with taking the bearings of a few known points. Happily Commander Pullen had made a special chart of Beechey Island in 1854, which was now of great use to us. Meanwhile the lieutenant took the opportunity of exploring the nature of the country, and collecting a great number of fossils.

Chapter II.

Northumberland House is the name given to a building erected on Beechey Island by Pullen in the autumn of 1852. It was intended to contain provisions

and equipment for Sir Edward Belcher's squadron, which was to search for Franklin. On the return of this squadron, the house and its contents were left behind as a depôt for Franklin, in case he should pass the island. Three boats of different construction were also left behind. Sir Leopold M'Clintock visited the place on his voyage of exploration with the "Fox" in 1858. Even at that time the depôt had begun to spoil; and when Sir Allan Young arrived there in 1878 with the "Pandora," it had been practically destroyed by bears. No wonder, then, that in 1903 we found the whole completely ruined. We took away with us the last remains of coal and a small quantity of sole leather, which was very acceptable. Although exposed for so many years to wind and weather, the leather was still quite good, and was even preferred to the Expedition's new "best American sole leather." But the fate of that depôt, it seemed to me, ought to be a warning to Arctic travellers who rely upon depôts 50 years old.

The marble slab erected by M'Clintock on behalf of Lady Franklin, to the memory of her husband and his companions and men, was found in order. It lay where it was placed in 1858, at the foot of the Belcher column, set up in memory of those who perished in Belcher's Expedition. In the same column is also placed a little memorial tablet to the French lieutenant, Bellot, who was drowned in these regions. All of these things we found in excellent order, as also were the graves themselves; we re-erected the only gravestone that had fallen down.

The heaviness and sadness of death hang over Beechey Island. Here is neither life nor vegetation. There is scarcely any water. When two men, after great trouble, at last found water to fill our tanks, and took it with them in one of our canvas boats, the boat sank and the water was lost.

THE " FRANKLIN, BELLOT, AND BELCHER " MEMORIAL, BEECHEY ISLAND.

A walk to the summit of the island gave us a fairly good view, though not as good as might be wished, owing to the continuance of the fog. We, however, got a peep a few miles out now and then. The sea is free from ice on all sides ; there is not a block to be seen anywhere. But what is that ? All at once the entrance is filled up with a heavy, white mass. It looks like a continuous mass of new ice emerging from the water, " pancake ice," as we call it. Our glasses are brought to bear on the phenomenon—the mass is seen to move. " Hang it all, boy, that fellow Morten should have been here ! " is the joyful exclamation of our fishermen at the sight of the mighty shoal of white fish now approaching.

On August 24th, about noon, we completed our magnetic observations. We had pitched our tent on the bank of a dried-up river bed. The spot was marked with tub staves, well rammed in, and large stones, so that a possible future observer will, it is to be hoped, have no difficulty in identifying it.

We once more all assembled at the old Franklin depôt and went carefully through everything, to see if there was still anything we might have any use for.

Several of the members of the expedition had set their fancy on an old hand-cart, and were anxious that we should take it with us. On being asked whether they would take it into their bunks, they gave in. Of course, they understood that we had no room. But the smith had made a discovery which sent him into the wildest raptures, a very ancient anvil. To dissuade him from taking it with him was impossible. The expedition would simply go to the bottom if we did not have the anvil with us. But we never had any use for it.

Chapter II.

We deposited an account of our progress, up to this point, in a tin case, and hung it up in the most conspicuous place, above the Bellot tablet on the Belcher column. Then we rowed to the ship and went aboard, all well satisfied with the sojourn at Beechey, and only wishing to get further ahead.

CHAPTER III.

In Virgin Water.

WITH the departure from Beechey, a new chapter opened in our expedition. We now knew the course we were to take ; the die was cast, and we only had to push on and make headway. Our voyage now assumed a new character. Hitherto we had been sailing in safe and known waters, where many others had preceeded us. Now we were making our way through waters never sailed in, save, possibly, by a couple of vessels, and were hoping to reach still farther where no keel had ever ploughed. We were very sanguine. In fact, I may almost say we felt certain that we should make our way through, having been fortunate enough to get thus far already. The ice conditions had been unusually favourable. We had made headway with ease, and almost without hindrance, where our predecessors had had to endure the most terrible struggles against ice and storms. As far as we could judge, the year 1903 must have been a very favourable one as regards ice. We set out from Beechey Island at 1 P.M. on August 24th, heading for Limestone Island at the entrance to Peel Sound. The compass—a floating compass by E. S. Ritchie, of

Chapter III.

Boston—proved most excellent. Owing to the nearness of the Magnetic Pole, it commenced, of course, to be somewhat slow in turning, but our further progress proved that it was quite serviceable ; for as soon as we got into Barrow Strait we were caught in a fog which hung heavy and dense till the 26th, and when it cleared we sighted the land around Peel Sound.

We encountered no ice with the exception of a few narrow strips of old sound ice, carried by the wash. Of large Polar ice we saw absolutely nothing. Between Sherard Head, on Prince of Wales's Land, and Cape Court, on North Somerset, we encountered the first large accumulation of ice. Having the sun in our eyes, we took it, in the mirror-like glitter of the calm sea, to be a compact mass of ice extending from shore to shore. It seemed evident to me that we had now reached the point whence our predecessors had been compelled to return— the border of solid unbroken ice. Happily we were mistaken, as, in fact, we were several times afterwards, under similar circumstances. With the sun right on the glassy surface of a sea with pieces of ice scattered over, these may easily present the appearance of one solid, continuous mass. This optical illusion is also enhanced by the "ice blink" constantly occurring in the Arctic Sea. This ice blink magnifies and exaggerates a small block of ice to such an extent that it looks like an iceberg ; especially when looking at it through a telescope at short range you may easily imagine you are facing a huge ice-pack. But on the Arctic Sea you can never rely on what you fancy you "see," however distinct it may appear. Certainty can only be acquired by actual contact.

As we drew nearer, the sparkling pieces of ice and the bright water seemed to part. It was the Nordlanders, Lund and Hansen, who first discovered our mistake. Added to their long experience in Arctic navigation, they had the advantage over us in their greater practice in the use of the telescope. The "mass of ice" proved to be simply old drifted out fjord ice which was quite loose. Between the ice and the land, on either side, there were large and perfectly clear channels, through which we passed easily and unimpeded. A large seal that lay basking in the sun on the ice paid for his indulgence with his life. The entire accumulation of ice was not very extensive. We were soon out again in open water, having escaped with nothing worse than a fright.

By 9 o'clock at night we were off Prescott Island in Franklin Strait. This island became a landmark on our voyage. The needle of the compass, which had been gradually losing its capacity for self-adjustment, now absolutely declined to act. We were thus reduced to steering by the stars, like our forefathers the Vikings. This mode of navigation is of doubtful security even in ordinary waters, but it is worse here, where the sky, for two-thirds of the time, is veiled in impenetrable fog. However, we were lucky enough to start in clear weather. Outside the promontories, some pieces of ice had accumulated ; otherwise the sea was free from ice. Next day we had a good lesson in our new mode of navigation, as clear weather alternated with fog all day long. I was walking up and down on deck in the afternoon, enjoying the sunshine whenever it broke through the fog. For the sake of my comrades, I maintained a calm demeanour as usual, but in reality I was inwardly much agitated. We were now fast approaching the De la Roquette Islands; they were already in sight. This was the point that Sir Allen Young reached with the " Pandora " in 1875, but here he encountered an invincible barrier of ice. Were we and the " Gjöa " to meet the same fate? Then, as I walked, I felt something like an irregular lurching motion, and I stopped in surprise. The sea all around was smooth and calm, and, annoyed at myself, I dismissed the nervousness from my mind. I continued my walk and there it was again! A sensation, as though, in stepping out, my foot touched the deck sooner than it should have done, according to my calculation. I leaned over the rail and gazed at the surface of the sea, but it was as calm and smooth as ever. I continued my promenade, but had not gone many steps before the sensation came again, and this time so distinctly that I could not be mistaken; there was a slight irregular motion in the ship. I would not have sold this slight motion for any amount of money. It was a swell under the boat, a swell—a message from the open sea. The water to the south was open, the impenetrable wall of ice was not there.

I cast my eyes over our little " Gjöa " from stem to stern, from the deck to the mast-top, and smiled.

In Virgin Water.

Would the "Gjöa" victoriously carry us all, and the flag of our native land, in spite of scornful predictions, over waters which had been long ago abandoned as hopeless? Soon the swell became more perceptible, and high glee shone on all our faces.

When I awoke at 1.30 next morning—it still amazes me that I could go to my berth, and sleep like a top into the bargain, that night—the swell had become so heavy that I had to sit down to put my clothes on. I had never liked a swell; there is something very uncomfortable about it, with its memories of nausea and headache, dating back to my earliest days of seafaring life. But this swell, at this place and time—it was not a delight, it was a rapture that filled me to the soul. When I came on deck it was rather dark, but on our beam, not far off, we could faintly discern the outlines of the De la Roquette Islands. And now we had reached the critical point, the "Gjöa" was heading into virgin waters. Now, it seemed we had really commenced *our* task in earnest.

The next doubtful point was Bellot Strait, where M'Clintock lay for two years waiting for a chance to get through. But the fairly heavy swell indicated an open sea for many miles to the south, and as Bellot Strait was not far ahead, our anxiety was not very great. At 8 A.M. we passed through the strait. The only thing we met was a very narrow strip of broken land ice. The strait itself was densely fog bound. Outside the sea was clear. As was to be expected, the swell was followed by a southerly breeze, and we toiled ahead rather slowly. At

Chapter III.

5.30 P.M. we met a quantity of ice off Cape Maguire, a fairly broad strip of loose ice. Beyond this we could see clear water. However, the fog settled down as thick as a wall just as we were about to make for the ice, and enveloped everything in its grey darkness. I decided to put back along the shore and wait till the fog lifted. The night was getting dark, and without a compass as we were, we ran the risk of getting into difficulties that might be pretty serious. So we lay to; but in the darkness of the night we felt many a heavy bump from the ice, and on the whole were far from comfortable. This was the first real drift ice that we met in the strait. Presumably it comes from M'Clintock Channel.

At dawn, 4 A.M., there was a slight break in the fog, but only for a moment. However, it enabled us to study the nature and appearance of the ice; and with the light wind blowing, to give us the direction, we proceeded merrily with the engine at full speed until 2 P.M. Then the fog cleared, and, bathed in glorious sunshine, the Tasmania Islands lay ahead of us. Thanks to the slight wind, which held out loyally all the time, we made satisfactory reckoning. The sun is certainly an excellent compass, but then it was rarely to be seen.

Hitherto the land along which we sailed had presented a mild and genial aspect, with luxuriant vegetation, but Tasmania Islands looked stern and bare. For once we were now favoured with a good wind; with the breeze on our beam a few points abaft, all sail set, and with the engine working at full power, we went splendidly towards

James Ross Strait. There was ice to the west, but along the land to the south appearances were favourable. I will reproduce here, verbatim, the entries in my journal for the following two days :—

"August 30th, Sunday. Made a somewhat faulty course last night, in the gloom and darkness, and became entangled in a large, tightly packed body of drift ice. It took us a couple of hours or so after daybreak to get out of the ice and into the channel. The coast water is very sharply defined here on Boothia. Presumably, the tide keeps the coast waters free from ice. Kept our course along the coast all day, and, according to dead reckoning, should have been near Cape Adelaide—the Magnetic North Pole of James Ross—about noon. Dull weather prevented us from discerning land. Our only means of guidance, the wind, baffled us again and again, as it was very variable. We have had a northerly breeze lately, and made good headway. The barometer fell a good deal to-day. It is raining, freshening up, and now at 9 P.M. it is pitch dark. It is no easy matter to navigate under these conditions, but still we can manage. We are now in the channel, laying to for the night. The land has quite altered its character since we left Tasmania Islands. It varies from high granite to low limestone.

"August 31st.—Last night there was a sudden marked fall in the barometer. The wind, which stood along the land from the northern side, freshened quickly, and rain began to fall. We lay to at 9 P.M. At midnight we had to reef sails as there was still a fresh breeze. The sea

Chapter III.

rose quickly and, strangely enough, as we neared the
Magnetic Pole, one or two of the expedition became
seasick. At 3 A.M. we made full sail again. The wind
had lulled a little, still the fog was fairly thick. We
kept close to the wind on the side where we supposed
the land lay. At 3.30 the fog lifted for a moment, and
we sighted a small island a little to leeward. Icebergs
and highly-piled pack-ice soon showed me that this island
was lying out towards the ocean itself. It was presumably the most northerly of the Beaufort group. We were
sailing closely to the wind, as we supposed to the south.
It subsequently appeared that the wind had veered to the
east, and this had caused us to drift a good deal towards
west. The fog parted several times but we saw nothing
of the land. At 8 A.M. I retired to my berth. We continued to keep close to the wind, bearing south and
intending to make for Matty Island. At 11 o'clock
I was awakened by a violent shock and was on deck in
an instant. We were aground just off a very low island,
which on further observation proved to be the southern-
most of the Beaufort Islands. The vessel had struck
amidships on a bank. We set all sail and started the
engine at full speed and threw out the kedge. After
awhile the vessel got off by means of the sails and the
engine, as we had not yet commenced to haul on the
kedge. It seemed we had got on a projecting shoal.
The vessel struck very hard several times, and some
splinters of the false keel floated up. The pumps were
sounded, but all was in order. After we got off we bore

eastward towards Boothia, keeping close to the wind, which had veered towards the south. Weather keeping fairly clear.

" At 4 P.M. we approached something which we took to be an island. The chart, in fact, showed a very small island here, but what faced us, as far as we could judge, was an island of very considerable extent and very flat. However, the chart proved to be at fault, and this long stretch of land running from north to south was, as a matter of fact, no island at all, but part of the mainland. I suppose that James Ross, when putting this down as a small island, did so at a time when everything was covered with snow, except a small eminence on the northern part of this projecting lowland. It appears that this flat coast bends, at its southern point, to the west, and almost joins one of the low-lying islands of the Beaufort group. We are now lying at anchor for the night under the lee of the land, in six fathoms of water. It is so dark at night that we can discern nothing ; and when, added to this, your course is an unknown one it is not surprising that the gravest errors frequently result. When we heave anchor again, early to-morrow, we shall be better able to see the second land point."

Here ends my journal entry for the day. From this short extract it will be evident to most people that navigation in the waters about the Magnetic Pole is by no means without its discomforts.

I was sitting at night entering the day's events in my journal, when I heard a shriek—a terrific shriek—which thrilled me to the very marrow—something extraordinary had happened. In a moment all hands were on deck. In the pitch-dark night, which luckily was perfectly calm, a mighty flame, with thick suffocating smoke was leaping up from the engine-room sky-light. A fire had broken out in the engine-room, right among the tanks holding 2,200 gallons of petroleum. We all knew what would happen if the tanks got heated : the " Gjöa " and everything on board would be blown to atoms like an exploded bomb. We all flew in frantic haste. One man rushed down to the engine-room to assist Wiik, who had stuck to his post from the outbreak of the fire. Our two fire-extinguishing appliances which were always ready for use, were first brought into play, and we pumped water on that fire for dear life. In an incredibly short time we had mastered it. It had broken out in the cleaning waste that was lying saturated with petroleum on the tanks. The next morning on clearing up the engine-room we found that it was no chance, but prompt discharge of duty, that had saved us all from certain destruction. Shortly before the fire broke out, Ristvedt had reported to me that one of the full petroleum tanks in the engine-room was leaking. I bade him draw the petroleum from that tank into one of the empty ones, immediately. This order was promptly carried out. On clearing up the engine-room after the fire, we found that the tap of the emptied tank had been wrenched right off during the struggle with the fire. Had my order not been carried out promptly over 100 gallons of petroleum

THE FIRE.

would have spurted out into the burning engine-room. I need not enlarge upon what would have been the inevitable sequel. But I hold up the man who so promptly obeyed orders as a shining example.

At 4 A.M. next morning we proceeded southwards along the coast. Elongated low islands with far-projecting shallows extended along this part of the coast. The weather was dark and as the wind was blowing a fairly stiff gale from abaft, the outlook was most uncertain. As the barometer was falling and the wind still freshening, I decided to seek shelter under, the lee of one of these islands and anchor there, to await fair weather. But these islands were so surrounded by banks that it was hopeless to get to leeward of them without grounding. I decided to make for the Matty side and seek a harbour there. A strong gale was now blowing. The soundings began to get deeper after we had put off from the coast, but we no sooner got ten fathoms than it began to get shallow again towards Matty Island. The sea was choppy and rough on the banks. At 11 A.M. I anchored in five fathoms of water to leeward of a low island, probably one of the Beverly Islands north of Matty Island. The gale increased steadily and was accompanied by heavy sleet. This was, indeed, glorious navigation.

During the night the wind slackened, and at 4 A.M. we weighed anchor and proceeded. The weather was tolerably clear, and the wind which had veered round to the west, was just the right strength for us. It was my turn at the wheel and I took my stand on the poop so as

Chapter III.

to have the best possible look-out. Lund and Ristvedt were busy stretching the mainsail. To leeward of us lay a low island with fairly extensive banks projecting to the east. We had seen this shoal from our anchorage so I knew how to steer to get clear of it. It was, therefore, an unpleasant surprise when we ran aground, although I had steered well out. We got off again immediately and I put the helm hard to starboard to sheer off from the bank, as it seemed to me that in spite of my reckoning, we had got in among the shallows jutting out from the islands. This, however, was a mistake ; as the shoal where we grounded was situated further to the south and west. Shortly after, we struck again, got off, and grounded again, this time for good. The engine was stopped, as also the work of setting sails. I rushed at once to the crow's nest. The weather was clear and I could see quite well. The bank we had grounded on was a large submerged reef, branching out in all directions. It extended to the west towards Boothia, as far as I could see. The land right to leeward was probably Matty Island.

It was 6 A.M. when we grounded. We immediately launched a boat to take soundings and ascertain the best way to get off again. The shortest way was aft. But as the two banks on which we had already struck lay higher in the water than the reef, on which we stood, the prospect of getting back over them was very slight. We were therefore obliged to try forward, to the south. The soundings gave us little hope. The reef shallowed up in that direction, and had not more than a fathom of water upon it in the shallowest part. Taking the shortest way ahead, the distance across the reef was about 220 yards. With a few tons of ballast the " Gjöa " had a draft of six feet. Loaded as she was, she drew 10 feet 2 inches. The prospect of getting across was therefore not brilliant, but we had no choice. We were compelled to lighten the vessel as much as possible. First we threw overboard 25 of our heaviest cases. They contained dog's pemmican, and weighed nearly 4 cwt. each. Then we threw out all the other cases of the deck-cargo on one side, to get the vessel to heel over as much as possible.

At 8 A.M. the current set to the north and the water fell one foot. We had grounded at high tide. We now made all preparations for the next high tide. The kedge anchor was put out, and every manœuvre was tried to make the vessel heel over. The weather continued fine and calm, with sunshine ; in other words, it was just the sort of day when we could have made good headway in these waters. Yet here we lay, and could not move an inch. However, we waited and trusted to the high tide. Our " observer " availed himself of the favourable opportunity to take our bearings. We were near a little island to the north of Matty Island. High tide was at about 7 ·P.M. But in spite of all preparations and all our exertions we could not get the vessel to move an inch forward. When darkness set in about 8 o'clock at night we had to give up for the day.

Chapter III.

When I came on deck at 2 A.M. next morning it was blowing fresh from the north. At 3 A.M. the vessel began to move, as if in convulsions. I had all hands called up so as to be ready to avail ourselves of any chance that might present itself. The north wind freshened to a gale, accompanied by sleet. We hove on the kedge, time after time, but to no purpose. The vessel pitched violently. I took counsel with my comrades, as I always did in critical situations, and we decided, as a last resource, to try and get her off with the sails. The spray was dashing over the ship, and the wind came in gusts, howling through the rigging, but we struggled and toiled and got the sails set. Then we commenced a method of sailing not one of us is ever likely to forget even should he attain the age of Methusaleh. The mighty press of sail and the high choppy sea, combined, had the effect of lifting the vessel up, and pitching her forward again among the rocks, so that we expected every moment to see her planks scattered on the sea. The false keel was splintered, and floated up. All we could do was to watch the course of events and calmly await the issue. As a matter of fact, I cannot say I did feel calm, as I stood in the rigging and followed the dance from one rock to another. I stood there with the bitterest self-reproach. If I had set a watch in the crow's nest, this would never have happened, because he would have observed the reef a long way off and reported it. Was my carelessness to wreck our whole undertaking, which had begun so

THE "GJÖA" AGROUND.

auspiciously ? Should we, who had got so much further than anyone before us—we who had so fortunately cleared parts of the passage universally regarded as the most difficult—should we now be compelled to stop and turn back crestfallen ? Turn back, yes ! that might yet be the question. If the vessel broke up, what then ? I had to hold fast with all my strength whenever the vessel, after being lifted, pitched down on to the rocks, or I should have been flung into the sea. Supposing she were broken up. There was a very good prospect of it. The water on the reef got shallower, and I noted how the sea broke on the outer edge. It looked as if the raging north wind meant to carry us just to that bitter end. The sails were as taut as drumheads, the rigging trembled, and I expected it to go overboard every minute. We were steadily nearing the shallowest part of the reef, and sharper and sharper grew the lash of the spray over the vessel.

I thought it almost impossible the ship could hold together if she could get on the outer edge of the reef, which, in fact, was almost lying dry. There was still time to let down a boat and load it with the most indispensable necessaries. I stood up there, in the most terrible agony, struggling for a decision. On me rested every responsibility, and the moment came when I had to make my choice—to abandon the " Gjöa," take to the boats, and let her be smashed up, or to dare the worst, and perchance go to meet death with all souls on board.

Chapter III.

I slid as quickly as I could down one of the back-stays on to the deck. "We will clear the boats and load them with provisions, rifles, and ammunition." Then Lund, who stood nearest, asked whether we might not make a last attempt by casting the remainder of the deck cargo overboard. This was, in fact, my own secret ardent desire, to which I had not dared to yield, for the sake of the others. Now, all with one accord agreed with Lund, and hey-presto! we went for the deck cargo. We set to in pairs, and cases of 4 cwt. were flung over the rail like trusses of hay. This done, up I climbed into the rigging again. There was not more than a boat's length between us and the shallowest part. The spray and sleet were washing over the vessel, the mast trembled, and the "Gjöa" seemed to pull herself together for a last final leap. She was lifted up high and flung bodily on to the bare rocks, bump, bump—with terrific force . . . . In my distress I sent up (I honestly confess it) an ardent prayer to the Almighty. Yet another thump, worse than ever, then one more, and we slid off.

I flew up to the top; not a moment was to be lost; everything now depended on our finding a way out among all the shoals which were lying close around us. Lieutenant Hansen stood at the wheel, cool and collected, a splendid fellow. And now he called out: "There is something wrong with the rudder, it will not steer." Should this, after all, be the end, should we drift down on the island there on our lea? Then the boat pitched once more over a crest, and I heard the glad shout: "The rudder is all right again."

A most wonderful thing had happened, the first shock had lifted the rudder so that it rested with the pintles on the mountings. But the last shock had brought it back into its place. It was a rare thing to see any frantic enthusiasm on board the "Gjöa"; we were all pretty quiet and cool by nature. But this time the jubilation could not be controlled and it burst out unrestrained.

The manœuvres that followed were far from agreeable. The banks lay all round us, and the vessel would not answer the helm as well as she usually did. We were drenched to the skin, and our teeth chattered with cold. The lead-line was brought into requisition and from that hour the "Gjöa" did not make another quarter of a mile of the North West Passage without one man aloft and another plying the lead. We had been taught one lesson, and we did not want another of the same kind.

Under sail and engine we stood over towards Boothia Felix, where we soon found deeper water. At noon we anchored to leeward of Cape Christian Frederik in five fathoms of water. A strong breeze was blowing from the north-east. We dropped both anchors at the same time, one with 30, and the other with 45 fathoms of chain. We had to make various repairs after the stranding, and besides, we were all fairly worn out after our toil and the severe mental strain.

In the afternoon some of us rowed to the shore to inspect it more closely and deposit a report in a cairn.

Chapter III.

I had arranged with Nansen how these cairns were to be built. They were to be erected on the most prominent points, and always in couples. The report was to be deposited in the larger one, and the smaller one was to be built 13 feet due north of the other. Should it become necessary to send out an expedition after us, they would be able to recognise our cairns from a considerable distance.

The geologist took his observations and collected a quantity of fossils. The sportsmen made an excursion inland and saw several reindeer. I myself sauntered round and explored some old rings where Eskimo tents had stood. There were many of these rings. When it was getting dark we rowed back to the vessel. The wind was slack, blowing from the land. When lying at anchor one man was always on the watch, the rest went to sleep.

At 11 o'clock the watch came and reported that a stiff wind was blowing from the south. When I came on deck it looked uncanny. It was completely dark, and there was a stiff sea breeze blowing. We had no option ; it was impossible to leave our anchorage as the water was so full of shallows. We paid out all the chain and hoped for the best. All hands were called up and seeing the position we were in we made everything clear in case we should run aground. We expected the anchor chains to part every moment owing to the heavy choppy sea and the force of the gale. The anchorage presented a hard bottom but luckily one anchor had caught in the cavity of a rock. We filled the flat-bottomed boat and the canvas boats with provisions and other necessaries. Each man had his task assigned to him, and we were ready should the chains snap. The engine was kept working full steam ahead, to relieve the strain on the anchors. Fortunately the chains held, but there we lay for five days and nights in terror, while the gale boxed the compass. It was not until 4 A.M. on the 8th that we were able to weigh anchor. A fresh wind was then blowing from the north-west.

The charts, prepared at a season of the year when the snow covering everything misled the draughtsman, were no more use here than they are in any other parts of these channels. It was impossible to work by them. When off Dundas Islands we lost sight of land in the thickly falling snow. This was owing to our having to stand further out on suddenly getting ten fathoms with the lead just after getting no bottom. We were probably crossing a ledge extending from Dundas Islands. Near Cape Christian Frederik the sea bottom changes from rock to clay, and as the colour turns to a light green it is difficult to detect the shoals.

At 3 P.M. we saw land right ahead, and I decided to bear down on it and seek a shelter for the night. The land lay very low, and extended southwards in pro- jecting points. I took the point I was heading for to be De la Guiche Point on the American mainland. But when we were still some distance from the coast the soundings decreased to four fathoms, and having in a clear interval sighted land on the opposite shore, we veered round and headed across the straits, hoping to find a refuge there. The land there was elevated, and we took it to be Mount Matheson on King William Land. By this time it was 5.30 P.M., and the prospect of getting there before dark was slight. The coast we had left terminated in a low point jutting out to the south-west, where the charts marked Cape Colville. If this was correct, Stanley Island, which forms the eastern shore of Rae Straits, ought to be in sight, but it was not. At 6 P.M. we got near three very low skerries. The current set strongly to the south, and threatened to take us on these. But we had cleared them when the wind freshened. Darkness fell before we sighted the high land again, and we had to shorten sail. The engine was kept going to prevent our making much leeway in the strong current. We took soundings all through the night.

According to our reckoning we should, during the night, have worked our way to the coast of King William Land. Our surprise was, therefore, great when, as soon as it was daylight, we found we were off the flat coast which we had, on the previous day, taken for Guiche Point. The current had carried us away in spite of sails and engine, and drifted us right in the opposite direction. We again set our course for the high land, and an hour's sailing brought us in sight of it ; and as we also sighted the three small skerries, our position was clear. The high land must have been

In Virgin Water.

Mount Matheson and the skerries were the Stanley Islands. In other words, we were in the middle of Rae Straits. To our pleasant surprise the soundings showed no bottom, and as we neared King William Land the weather became quite clear, with a fresh breeze from the north. From Mount Matheson a long stretch of low-lying land extended in a south-easterly direction ; this being a terminal point of King William Land, we christened it Point Luigi d'Abruzzi, in memory of the Duke d'Abruzzi. A number of small islands off the coast were not charted ; these we called Eivind Astrup's Islands. These islands and the Duke d'Abruzzi's Point form a good safe entrance to Simpson Strait. From the point to Neumayer's Peninsula there is a very wide bay—Schwatka Bay—extending about 10 miles inland. A fresh northerly wind was blowing from this bay. When we were off Betzold Point I decided to stand in for Pettersen's Bay and anchor there for the night. This proved to be a very lucky hit. There was perfectly smooth water under the lee, and although we had to tack up the bay, we managed it very quickly. From the deck there was nothing particular to be seen except the large wide bay. But Hansen, who was on the look-out aloft, saw more than we did. He suddenly called out : " I see the finest little harbour in the world." I climbed up to him, and true enough I saw a small harbour quite sheltered from the wind, a veritable haven of rest for us weary travellers. We afterwards christened it " Gjöahavn." We anchored outside in four fathoms of

Chapter III.

water. The wind was blowing strong from the narrow entrance, but we would not venture any nearer till we had taken soundings and surveyed the shore.

To the westward Simpson Strait appeared, quite free from ice. The North West Passage was therefore open to us. But our first and foremost task was to obtain exact data as to the Magnetic North Pole, and so the Passage, being of less importance, had to be left in abeyance.

As soon as I saw Gjöahavn, I decided to choose it for our winter quarters. It was evident that the autumn storms had set in in earnest, and I knew the waters further west were very shallow. Before deciding definitely upon this course, I intended to explore the harbour in a boat. The Magnetic Pole, as shown by our observations, appeared to be situated somewhere in the neighbourhood of its old position, and as Gjöahavn was about 90 miles from that locality, it should, according to the dicta of scientific men, be particularly suited for a fixed magnetic station. If we were to get our observatories built, and everything else set in order for wintering here, we should have to bestir ourselves. We had, moreover, been working very hard during the last few weeks, and needed a rest. As regards myself, at any rate, I confess that I wanted breathing time. So why look further west for a harbour, which possibly we should not find. Had the completion of the North West Passage been our chief object, it would have been a different matter, and nothing would have prevented us from going further on,

THE " GJÖA " AT ANCHOR IN GJÖAHAVN (SUMMER OF 1904).

In Virgin Water.

At 6 P.M. I rowed into the harbour with Lieutenant Hansen and Lund. The entrance was not wide. At the narrowest part there was scarcely room for two vessels to pass each other, but the soundings, averaging six fathoms, gave ample depth. The harbour itself was all that could be desired. The narrow entrance would prevent the intrusion of large masses of ice, and the inner basin was so small that no wind could trouble us there from whatever quarter it blew. The shore around the harbour was very low, sandy moss-clad ground, gradually rising to a height of about 160 feet. Two little streams provided fresh water; if these dried up, as seemed probable, there was, right on the crest of the ridge, a fairly large pond containing drinkable water. A number of cairns and tent circles showed that Eskimo had been there, but that, of course, might have been a long time ago. Fresh reindeer tracks gave hope of sport; there was not a trace of snow, and large stretches of moss were quite parched, showing that the summer had been very warm. The spot seemed eminently suitable for a magnetic station. There was no rocky land in any direction which, by the iron contained in it, might have a disturbing effect on the observations; the sand, of course, might be ferriferous, but there was little probability of that. The results of our investigations proved favourable in every respect, to the great joy of all on board.

The day after, Lund, Hansen, and Ristvedt went ashore to test the chances of sport. In the afternoon

Chapter III.

they returned with two reindeer calves and one doe. They had seen a large herd of reindeer and a quantity of birds. Our mouths watered when they told of large flocks of geese. For the rest, they said the country was an ideal one for reindeer, being flat, mossy, and abounding in streams and lakes. Eventually, on Saturday, September 12th, at 7.30 P.M., the north wind fell sufficiently to permit of our venturing in with the aid of the the engine. At 8.30 the "Gjöa" was anchored in Gjöahavn.

We had got thus far. A good deal of work had been done, and we had every reason to be pleased.

Before I continue my narrative, I think it appropriate, at this point, to explain as briefly as possible, the nature of terrestrial magnetism, and the use of our magnetic instruments.

The magnetic force of the earth manifests itself differently, as regards its direction and intensity at every single point of the earth's surface, and even at one and the same point, it is not constantly the same. It is subject to regular daily and annual variations, and more or less violent disturbances frequently arise. Lastly, slight gradual displacements manifest themselves from year to year, which continue at the same rate for long cycles of years.

All this has been found by observations, made in the course of generations all over the globe; partly during voyages and travels, partly at fixed stations. A close study of the material results of observations extant at the

GJÖA HAVEN AND ITS ENVIRONS

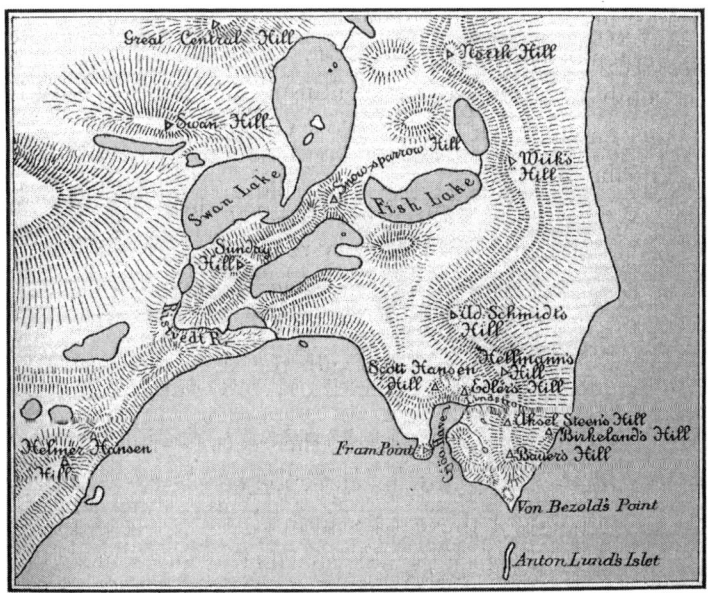

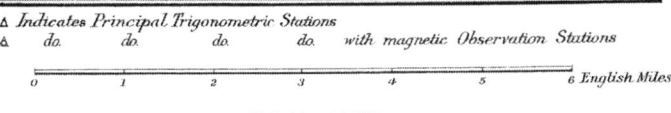

△ Indicates Principal Trigonometric Stations
▲   do.   do.   do.   do.   with magnetic Observation Stations

0   1   2   3   4   5   6 English Miles

time, led Gauss, the great German mathematician in the thirties of last century, to set up a theory on the connection existing between the various phenomena of terrestrial magnetism, and their varying manifestations at a given moment, according to the geographical latitude and longitude. Thereby it became possible to draw up three different charts, of which two indicated the direction of the magnetic force, and the third showed its intensity. The reason why two charts are required to show the direction of the force is that the direction must be shown both in relation to the geographical " North to South " line of the place, and in relation to the " horizontal plane " of the place. The direction of terrestrial magnetic force in relation to the " North to South " line can be observed with the aid of a compass needle, of which the end pointing towards the North is known as a rule to be directed a little to the east, or a little to the west of the true north. This variation is also called " deviation " or " declination." On Chart No. I, lines are drawn which show the direction of the compass needle at every point of the earth's surface. These lines, which are called " Magnetic Meridians," converge, as will be seen, in two points, viz., the Magnetic North Pole, near the North American Arctic Coast, and the Magnetic South Pole, situated on the Antarctic mainland. Each line indicates, as will be readily understood, the course which one would have to take if travelling straight ahead exactly in the direction indicated either by the north point or by the south point of the needle. If the former, we should in the end arrive at the Magnetic North Pole, if the latter, we should reach the Magnetic South Pole.

Chart No. II gives us an idea as to the direction of magnetic force in relation to the "horizontal plane" in various parts of the world. If we fit up a magnetic needle so that it can revolve on a horizontal axis passing though its centre of gravity (exactly like a grindstone), the needle will, of its own accord, assume a

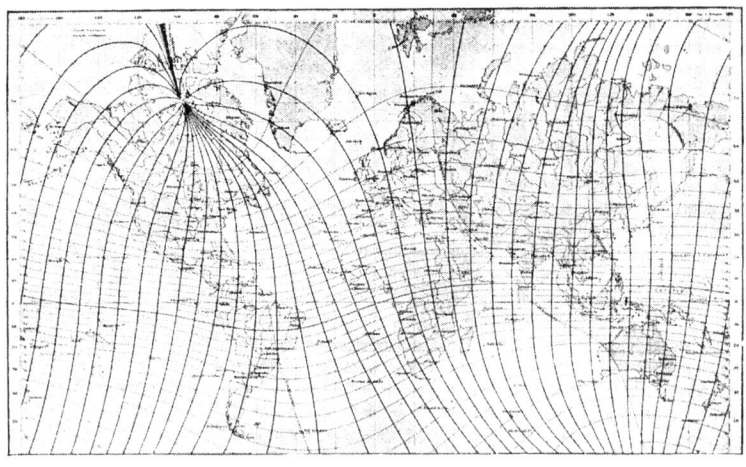

CHART I.—MAGNETIC MERIDIANS, 1885.

slanting position if its plane of rotation coincides with the direction indicated by the compass. Such an apparatus is called an "inclinatorium" or "dipping needle," and the angle which the dipping needle forms with the horizontal plane, is called the "magnetic inclination" for the respective locality. Here, in our regions, it is the north point of the needle which dips towards the earth, but in Australia it is the south point of the needle.

In Virgin Water.

At the Magnetic North Pole, the dipping needle will assume a vertical position, with its north point directed downwards ; at the Magnetic South Pole it will stand vertically with its south point downwards. Hence, in both places, the inclination is 90°, and it will decrease in proportion as we move further away from the poles. In a series of points in the tropics the inclination is 0°, that is to say, the dipping needle assumes an exactly hori-

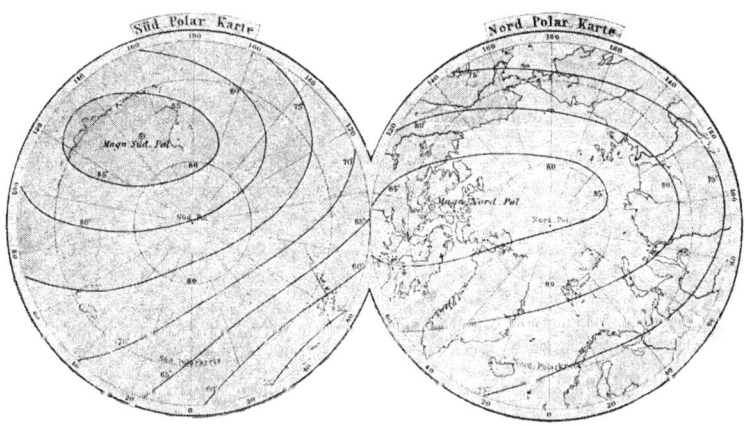

CHART II.—MAGNETIC INCLINATION LINES, 1885.

zontal position ; the imaginary line drawn through all these points is called the "magnetic equator." This is situated partly north and partly south of the geographical equator.

It will be understood that terrestrial magnetism acts with its full intensity in the direction indicated by the dipping needle, and the question may be raised, how great this intensity would be in any particular locality? In order to elucidate this, we will imagine the magnetic force to be split up into two components, one acting horizontally, and the other vertically. Evidently, it is the horizontal component of the force which causes the compass needle to point in a certain direction, and if we can determine this component of the "horizontal intensity" as it is called, and if, at the same time, we know the inclination, it is easy by a simple calculation to find the aggregate power—the "total intensity." For determining the horizontal intensity, two methods were employed, either each separately, or preferably, for checking purposes, both together. One method consists in fitting up a bar magnet by the side of a compass needle, at a stated distance from the latter, and observing how many degrees the needle is deflected from its original position. It is manifest that this deflection will be all the greater, the less the horizontal intensity is in the respective locality, and if we know the magnetic power of the bar-magnet used, we can calculate the horizontal intensity from the angle of deflection, and the distance between the bar-magnet and the needle.

The second method is based on the observation of the period of oscillation of a bar-magnet suspended on a thread so that it can turn in the horizontal plane. When the magnet is in a state of rest it points, under the influence of the horizontal intensity, in the same direction as the compass needle. If, now, it is brought out of the position of equilibrium by a slight tap, it will oscillate to and fro, and the greater the horizontal intensity the more quickly it will return to its state

of rest ; or in other words, the shorter will be the period
of each oscillation. If, now, we know the magnetic
power of the oscillating magnet, and observe how many
seconds it requires for each oscillation, we can calculate
the horizontal intensity.

Chart No. III gives us an idea of the value of the
horizontal intensity, expressed in so-called electric units,

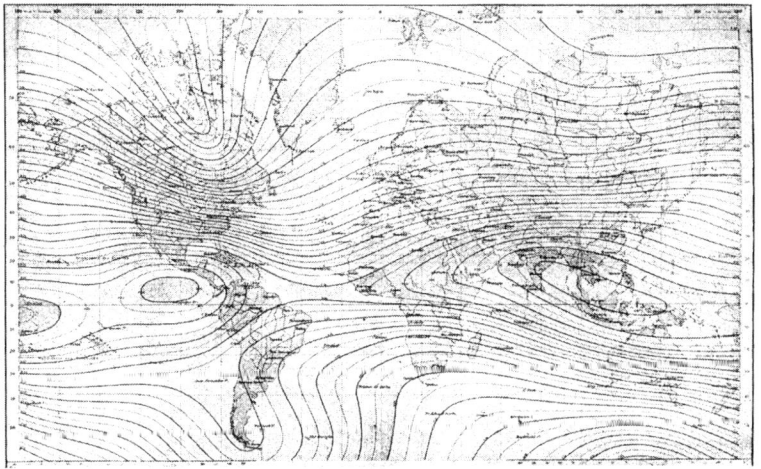

CHART III.—LINES THROUGH PLACES HAVING THE SAME HORIZONTAL INTENSITY.

for the various parts of the earth. Each line passes
respectively through all the places in which the horizontal intensity is identical. It will be seen that the
horizontal intensity decreases towards the magnetic
poles. This, in fact, is self-evident, because at the
poles, where the inclination is 90°, terrestrial magnetism
acts with its full force in a vertical direction downwards,
and therefore cannot exert any influence on the horizontal direction.

Chapter III.

However much the charts here shown may differ, they agree in this that the Magnetic North and South Poles are the cardinal points on the earth's surface, and it is evident that magnetic observations made exactly on

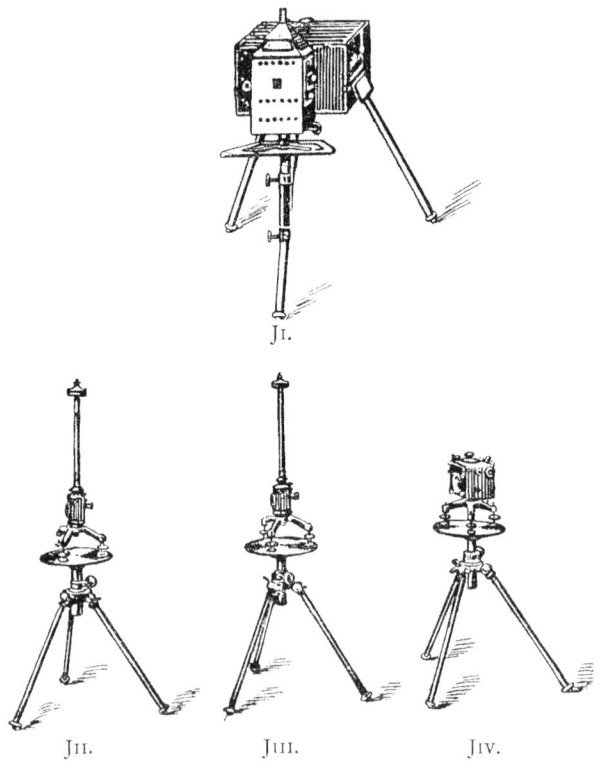

Jɪ.

Jɪɪ.   Jɪɪɪ.   Jɪᴠ.

SELF-REGISTERING MAGNETIC VARIATION APPARATUS.

these points, or in their immediate neighbourhood, must be of the greatest interest to the science of terrestrial magnetism. Gauss's theory does not by a long way explain all the problems presented by the terrestrial magnetic phenomena, but men of science are continually labouring to complete it, by collecting the most reliable and exhaustive observations which it is possible to obtain.

The labours of the "Gjöa" Expedition in this direction were intended to form a contribution towards these data. But the difficulties were no slight ones. The circumstance alone that, as we have seen, the "horizontal intensity" becomes infinitesimal in the neighbourhood of the Magnetic Pole, calls for extraordinary precautions in order to determine both this and the "variation." The equipment, by way of instruments, of the "Gjöa" had, in fact, been specially adapted to these conditions. The magnets, fourteen in number, to be used for determining the horizontal intensity, had been selected with great care in Potsdam before our departure. For determining the "inclination" there were three dipping needles of different construction, and for determining the "declination" we had two different instruments.

In addition to these, we had a set of self-registering variation apparatus (see illustration, p. 90), that is to say, three instruments fitted up on a firm base in a dark room, each containing a small magnetic needle, of which two were suspended upon a fine quartz thread, and the third pivoted on delicate bearings, so that the needle responded by its movements to the slightest fluctuations : the first, to those of the "declination" (J$_{II}$) ; the second, to those of the "horizontal intensity" (J$_{III}$) ; and the third, to those of the "inclination" (J$_{IV}$). Each needle was provided with a mirror, which reflected the light of

Chapter III.

a lamp upon a drum covered with sensitised paper (J1),
which, by means of clockwork, made one revolution in
24 hours. The arrangement was such as to cause the
ray of light reflected from each of the three needles to
strike the drum at varying heights and produce a little

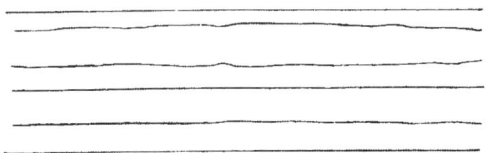

MAGNETIC VARIATION DIAGRAM FOR TWENTY-FOUR HOURS UNDER
QUIESCENT MAGNETIC CONDITIONS.

dark spot ; but as the drum with the paper was revolving,
each little spot steadily advanced on the paper and produced a continuous dark line. Thus when, at the end of
24 hours, the paper was removed, three dark, more or
less irregular lines had been produced, one above the

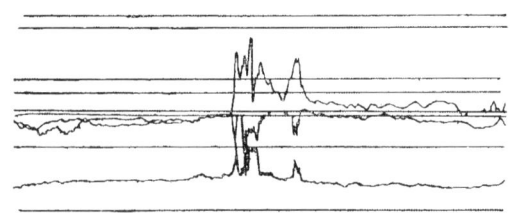

MAGNETIC VARIATION DIAGRAM FOR TWENTY-FOUR HOURS WITH
MAGNETIC DISTURBANCES.

other, on the paper, in addition to the three straight
horizontal direction lines.

The illustrations above show examples of the three
magnetic variation diagrams, for a period of 24 hours,
during quiescent magnetic conditions, and for a period of
24 hours during which magnetic disturbances arose.

In Virgin Water.

From what has been previously stated it will be readily understood that it would not do to select the Pole itself as the site of a fixed station for observation, even if we knew its exact situation beforehand, and if we could take it for granted that it remained immutably in one and the same spot. In accordance with Professor Adolphus Schmidt's advice, I therefore decided to establish the "base station," where the variation instruments were to be fitted up, so far away from the Pole that the inclination would be about 89°.

The day after my arrival I went ashore with my dipping needle to examine the magnetic conditions of the place. A series of observations gave an inclination of 89° 15', or about 90 nautical miles distance from the Pole itself. It could not have suited us better; we could not go any nearer.

On Monday, September 14th, at 5 A.M., we brought the vessel right up to the bank and berthed there, just as we should alongside a quay. We were thus ready to commence preparations for our proposed winter quarters. First came the turn of the dogs, who were all taken ashore in the flat-bottomed boat. Selecting a sheltered little valley, we drove into the sand two wooden stakes, stretched a rope between them, and tied the dogs to the rope. The dogs, of course, were highly affronted at being thus summarily ejected from the ship, but it was a great relief for us to get rid of them, as they were in the way, and gave no end of trouble there. After this "eviction" we constructed an aerial ropeway to facilitate

unloading, as it was my intention to carry all provisions
ashore, so as to make as much room as possible on board.

THE " GJÖA " DISCHARGING BY AERIAL ROPEWAY.

Lindström, also, was to have room in the hold for all his
cooking apparatus The ropeway consisted of a steel
cable stretched across from the middle of the mast to the

old coast-line, about 20 yards above the present one, where we had found a convenient storage place for the cases. Ashore, the rope was made fast to the kedge, which we had buried fully a yard deep in the sand, and hooked fast in the frozen subsoil. A pulley block travelled to and fro on the cable, with an "inhauler" both ashore and on board. The cases were hoisted out of the hold by means of a tackle hitched to the pulley-block and let go; they then travelled ashore quite smartly. Ristvedt and myself received the cases on shore; the others were busy with them on board. We landed them on a bed of planks. As soon as a case came on shore we knocked off the wooden lid, turned the case upside down, and lifted off the outer wooden case so that the inner tin-chest was laid bare. An exact sketch was prepared of the chests when placed in position, and the numbers and contents noted down so that we could at any time find what we wanted. The empty wooden cases were carefully collected and put aside to serve later on as building materials.

We worked from 5 A.M. to 6 P.M. The "eight hours day" was not yet introduced among us; that was to come later on. We had some foretaste of winter, in the form of snow and sleet, but we hoped that it would still keep off and give us time to get ready.

By the afternoon of the 17th the work of discharging was finished. We put up a sailcloth awning over the cases, and the whole thing looked very smart. As this provision store lay on rising ground, there was a possi-

Chapter III.

bility of moisture filtering in, although the soil was sandy and the water would, therefore, probably sink into the ground ; but for greater security, we dug a deep trench around the whole shed. After the provisions, the explosives came next in turn ; these were carried far in-shore and a little shed was built over them. Subsequently, our

BUILDING THE STORE HOUSE. GJOAHAVN, 1903.

clothing and all the goods which would not bear moisture, were brought over to the store-house, which, in fact, proved to be the driest place we had.

Then we began clearing up on board. First of all we set the hold in order. Then the galley, which was amidships, was unscrewed, taken to pieces, and set up again in the hold. Lindström was thus in command below, and had his kitchen there from September, 1903, to June, 1905. This work demanded the assistance of all hands. Afterwards, in order to accomplish as much as possible before the winter set in, we divided ourselves into two parties. First and foremost it was necessary to get our observatory erected, and to procure fresh meat for the winter. Reindeer had hitherto shown themselves very rarely in our neighbourhood. Lund and Hansen were, therefore, sent out on a boat trip to the little Island of Eta which lies in the middle of Simpson Strait, and where I knew from reports that reindeer used to come in autumn in large herds. On September 21st they went off in the " dory," provisioned for a fortnight. We who remained took the building operations in hand.

Wiik had meantime determined the magnetic meridian —the line of variation from the true " North to South " line—showing in what direction the " Magnetic variation " house, with the self-registering instruments, was to be set up. The outer cases of the provision chests, which were intended to serve as building material for this construction, were carefully examined to make sure that there were no iron nails in them. They were all made up to uniform size, and joined with copper nails, which would not exert any influence on the magnetic observations. We selected the building site on the crest of the ridge facing Simpson Strait. A foundation of stone was laid down and cemented over, to form the base on which the instruments were to rest. Then we built the house. Case after case was set up and filled with sand. This took 40 cases. Outside and inside, the house was covered with waterproof felt, and finally the whole was weighted down with sand ; around the whole building we dug a deep trench to carry off the water. It was Lieutenant Hansen and myself who had to do this job, and

BUILDING THE MAGNETIC VARIATION HOUSE. GJÖAHAVN, 1903.

we well remember it. More than once we stretched our aching backs, unaccustomed to this kind of work, and wished trench-digging to the deuce. But we managed it, and on September 26th the observatory was quite ready.

Lund and Hansen came back from their hunting expedition late that night. They had been lucky, and the boat was loaded with twenty carcasses of reindeer. At a distance barely twelve miles from the harbour, they had found a spot where there were large herds of reindeer, but the animals were very shy and difficult to get at. They put up a tent and hunted for several days. This place was near Booth Point, which later on became a familiar spot, as we met Eskimo who had their camp there. The Eskimo told us afterwards that they had seen our huntsmen, but dared not approach them for fear of the guns.

Hansen and Lund reported that the land we saw to the south of our station was not Ogle Point, as we thought, but an island. They had gone so far south west that they could see the island clear on all sides. It greatly surprised me to find an island here not charted by M'Clintock. But our huntsmen's statement was verified; presumably M'Clintock passed it in a fog.

On the 29th we commenced building the house in which Ristvedt and Wiik were to live. Sixty cases were required for this. It found a site on the same ridge on which the observatory was erected, 250 feet away, and with a commanding outlook on all sides. We also did sundry work on board. Double skylights were put in, oil-stoves were set up and the ventilation was improved. We made ourselves comfortable in the cabin, and it was very enjoyable after finishing the day's work to come into the cosy well-lighted room and have a good meal.

Chapter III.

We lounged about on those nights with an exquisite sense of comfort after the rough daily toil. We could not but feel that we had been very fortunate in every way, not least as regards meat, as the twenty reindeer carcases had been well quartered and hung. It was now so cold that there was no fear of it going bad. Finally, on the 29th, the whole vessel was covered in with sailcloth, and we were ready to stand the winter on board.

CHAPTER IV.

THE FIRST WINTER.

ON October 1st there was every appearance of winter. The sea water dashing against the boat in the north-easterly gale was immediately frozen, clothing the "Gjöa" in a solid armour of ice. The drifting snow swept into our eyes, and as it mingled with the sea water it formed a slush that spread over half the sea. This was the initial stage of the ice formation. As soon as the wind lulled the ice would be strong enough to bear.

Meanwhile the "Gjöa" had been driven right up on the shore by the storm during the night; our anchors had not had a sufficiently firm hold. This of itself did not make any difference either to the boat or to us. But if the formation of the ice was commencing in earnest, we could not leave the boat lying hard up on the beach, for fear of the tide; so as soon as the gale had abated next day, we brought her off, and the "Gjöa" took up her winter quarters about fifty yards from the shore. We did not care to take her further out, not wishing to be far from our cherished store tent. On the 3rd we were able to walk over the ice to the shore—that

Chapter IV.

is to say, those who were not too heavy on their feet. As we each had to begin our allotted tasks after breakfast, we all remained on deck.

On the crest of the ridge above the harbour, a herd of about fifty reindeer came in sight. They came in single file, light and graceful, a huge buck leading the van. He was a really splendid specimen, with enormous antlers, and long white shaggy hair hanging down from his throat; obviously, he was in undisputed command, and was on well-known ground, for he took the shortest way to the beach. It was evident they intended to inspect the newly-frozen ice and see whether it was strong enough to bear them, so that they could cross the Sound to the mainland. Their summer sojourn on King William Land, where they spend some months in peace and quiet, was now ended. Those endless moss tracts, with their thousand lakes, must be a veritable paradise for reindeer. A further advantage of the islands is that the wolf, their deadly enemy, curiously enough does not follow them from the mainland in the spring. That the reindeer do not prefer to remain during the winter in those peaceful regions is probably due to their being deterred by the rough climate, with its many severe storms.

The wind was blowing off the land, so that they could not get scent of us. We could therefore observe them at our leisure, as they crossed over the ridge and disappeared on the other side. But now was the time for our sportsmen. All work was discarded, the rifles were brought out, and we were off in pursuit. I myself am not really a sportsman. I cannot fancy shooting an animal just for amusement. I therefore left these shooting expeditions to my comrades, who were all passionate sportsmen, and I took upon myself the certainly less enjoyable but none the less necessary part of the business, of bringing the game in. Lieutenant Hansen, who was a born sportsman, exhibited great self-denial in joining the transport department. Unaccustomed as we were to driving with dogs, these meat transports were at first rather trying.

The chase in these regions is interesting, but by no means an easy matter. The reindeer is exceedingly shy, and the endless wastes do not afford the least cover. The sportsman must creep up to the game, like a serpent, and carefully watch the direction of the wind. If the quarry gets the slightest scent of the hunters, the whole herd scampers out of sight. It is easier to approach them when they are grazing, but if they have had their fill and lie down before the hunter can get within range, he must lie and wait patiently till they rise and resume grazing, which may not be for hours. The Eskimo are far superior to us white men in enduring this trial of patience. To them time is of no account, while booty is everything. They will pursue one reindeer from morning till night.

The ground about Gjöahavn was the most broken ground on the whole of King William Land, which, in fact, is not saying much ; therefore our sportsmen liked

Chapter IV.

best to hunt round the harbour. It so happened, in fact, that this autumn we could get as many reindeer, close by us, as we could do with. Herd after herd, often numbering several hundred head, passed by us and took their way to the sea. I am inclined to believe that this was due to the fact that the ice this year formed a different track from the usual one. When the reindeer came to their usual place for crossing over, on the narrowest part of Simpson Strait, near Eta Island, they found open water there. They therefore made their way along the coast to seek a passage elsewhere.

At 11 A.M. three of the huntsmen came back, having killed one buck, two does, and two calves. The fourth hunter arrived later. He had been close up to a large herd, but the mechanism of his gun had gone wrong just as he was about to fire. Such troublesome accidents frequently happened ; but, perhaps, the most annoying part of it was the evident incredulity with which these statements were invariably received by the others. The engineer was, without doubt, our keenest sportsman. He took every opportunity to go out hunting. But he was prepossessed by the fixed idea that, to be lucky, he must absolutely have a soft grey felt hat on his head. As he had lost all his own hats in the course of the voyage, he went first of all on the hunt for other people's hats. The result was excellent ; next Sunday our sportsman appeared with a particularly fine, brand-new grey felt hat. That the hat was mine did not seem to trouble him in the least. And I must admit that, with the run of luck he now had in sporting, he did credit to my hat. That very day he shot five ptarmigan, the first of the season.

One day now passed like the other, with the sole alternation of work and sport, and sport and work. A hunting record, which stood unbeaten during the whole of our stay, was the one made by Helmer Hansen on October 8th. He returned, after being out shooting for a short time, having shot thirteen reindeer, quite a creditable performance for a single rifle. However, this was too much for my small transport detachment. We had to requisition extra help, and the whole crew had to go into harness. While we were just getting ready for a start, a doe with two calves came walking down, quite unconcerned, towards the vessel. As soon as they came within range, they were received with a brisk fire, and all paid with their lives for their *sancta simplicitas*. Much snow had now fallen, and there was therefore good sledging. Thus it was comparatively easy, and did not take us long to bring the game in.

As the country about here is monotonous, and without prominent landmarks, it is sometimes difficult to find one's way. The other night we were out with the sledges to bring game in, in two parties, the lieutenant and myself forming one, and Ristvedt and Wiik the other. It was quite dark when Lieutenant Hansen and I returned on board, but the other two had not come in. They only arrived a couple of hours later, and had, in fact, been well on the way to complete the North

Chapter IV.

West Passage by sledge on their own account, as in the darkness they had passed the harbour without perceiving it, and gone on further west. When at last they suspected that they had missed their way, they left the sledges behind, and returned, following the shore.

The Krag-Jörgensen rifle, which we used, was a splendid weapon, but we had to use lead projectiles; steel projectiles were no use. A reindeer would run away as fast as ever even when several of these bullets had lodged in its body. In drifting snow we frequently had excellent shooting. Probably this was because the snowstorm blinded the eyes of the animals, and under such conditions it is nothing unusual to be able to get quite close to the game. It is astonishing how lean these animals were. With the exception of two bucks, which had a little fat on their bodies, they were all exceedingly lean. Possibly the reason is that the warm summer, with the blazing sunshine, had parched up the moss. I do not know, but it seems most likely, that their leanness was due to want of food. The following autumn, after a poor, cold summer, they were like well-fattened pigs.

The cook, who had volunteered to collect zoological material for the University, had already procured several fine specimens. Thus Ristvedt and Wiik had each enriched his collection with the skin of a splendid buck. A special wish for such reindeer bucks had been expressed by the University, and you may imagine the joy of our "superintendent," when our Nordlanders, with their usual skill, had stripped off the skin and hung it up for preliminary drying.

The dogs, which had hitherto been obliged to lodge under the open sky, now had a kennel built for them. This was dug out in a huge snowdrift. One of our flat-bottomed boats was laid over it as a roof, and there stood the finest kennel that could be desired. The whole structure was sprinkled over with sea water, so as to consolidate it into one compact body. It was divided into two compartments. One was assigned to the old "Fram" team, the other to the Godhavn team.

At the same time another most important job was carried out. A hole was cut in the ice on the starboard side, and a snow hut built over it. This hole was kept open during the whole winter, so that we might have a supply of water in case of fire. This establishment was called the "Fire Station," and Lund was appointed Chief of the Fire Brigade. It was not altogether a pleasant position to be Chief of the Fire Brigade in Gjöahavn. Every morning he had to go out to the hole and do what was needed to keep it open. When the ice had attained a thickness of about four yards, as was the case during the first winter, this was no easy task.

The snow which had fallen had now drifted together so compactly as to form an excellent building material. I therefore set to work, with the assistance of Lund and Hansen, to erect a building in which we could make the absolute magnetic observations during the winter. A site for this was chosen at a distance of 250 feet from the

Chapter IV.

"Variation House," and it was built in the direction of the magnetic meridian. We fetched the building material from a neighbouring valley, where the snow had been swept together in a hard mass in large quantities. The building was to be 26 feet long by 6 feet 6 inches wide by 6 feet high. The blocks were cut out of the snow with a saw. An idea of the compactness of the snow may be gained from the fact that these blocks weighed on an average 2 cwt. each. When we came to

OBSERVATORY FOR ABSOLUTE MAGNETIC OBSERVATIONS, 1904–1905.

put on the last tier, three men were required to hoist the blocks into position. A roof of thin transparent cloth was subsequently made up and put over the erection. We thus obtained an excellent building for the absolute magnetic observations.

As the cold was now setting in sharp, and made itself unpleasantly felt, we had to think of our personal outfit for the winter. Thanks to our luck in reindeer shooting, we possessed a large supply of excellent skins. The

The First Winter.

Lieutenant and I were constantly deliberating how to render these serviceable for underclothing. We had brought out with us a good supply of outer-clothing of reindeer skin, and had, therefore, no need to trouble about this part of our attire, but it would be splendid to have some nice, soft underclothing. We, therefore, selected all the calves' skins, took them down into the cabin, and set to work. Neither of us had the least idea how to go about it. We knew, indeed, that we ought to spread them out to dry, but as to whether the drying should be done by gentle heat, or by a quick fire, we had not the slightest notion. He looked at me, and I looked at him. However, we arrived at the conclusion that it would be best to stretch them out up under the cabin roof. As many skins as there was room for were extended overhead, and soon the cabin resembled a combined butcher's and tanner's establishment. We felt the skins every day, and when we thought they were sufficiently dry, we took them down and commenced dressing them. How we laboured! We both were anxious to obtain the best results. How far we might have succeeded is hard to tell; I believe that, after all, in the absence of outside assistance we should have contrived to botch up something by way of under-clothing, though not, perhaps, of first-class quality. But in the hour of need, help came before we had dreamed of it.

On October 17th, Ristvedt and Wiik had quite finished building their house. It was forthwith christened the "Magnet." It was not distinguished so much by its appearance as by its situation. It lay on the top of an eminence about 100 feet high, out towards the sea, affording a magnificent view over the whole of Simpson Strait. Thus nothing could happen without being observed from the "Magnet." If strangers arrived they would have to pass here in most cases. If a bear should turn up on the ice in the Strait he would

VILLA "MAGNET" DURING THE SUMMER. THE METEOROLOGICAL CASE IS SEEN CLOSE BY.

be seen from here. In short, the occupants of the "Magnet" commanded the whole country. The house was built, like the first, of cases filled with sand. It was not a palace, but we all were agreed that the two occupants were considerably more comfortable there than anyone on board. It only contained one room—bedroom and workroom combined. In one corner there was a large, broad bedstead, manufactured of boards of packing cases. They had found that one bed took up less room than two. They had also found that two in one bed could keep warm better than one, and in this one must admit they were right. All in all, the whole was fitted up in a most practical manner. Just by the bed there was a table with a bench on either side. The other half of the room they had divided so that Ristvedt had his work-bench on one side, and Wiik his "diagram table," where he produced the magnetic curves, on the other side. The floor was covered in with boards and reindeer skins. The house also had two windows, one looking out towards the sea, and the other towards the "Gjöa." Whenever occasion arose, the windows were covered up with earth and sand. As far as I know, this is the first time that packing cases have been used in the Polar regions as building material. If something to fill them with is available, I recommend them as superior to any other material ; if not, of course, it may be a different matter. Wiik and Ristvedt lived in the "Magnet" for nearly two years, and they would hardly have cared to change places with us on board. Lieutenant Hansen and I lived in the cabin together. Our quarters were very damp, and every night during the winter we had to chop large icebergs out of our bunks. Lund, Hansen, and Lindström resided in the forecastle. It was also somewhat damp there, but not so bad as in the cabin aft. The first winter we had the vessel entirely covered with snow, and the temperature did not then fall below freezing point in the fore-cabin. In the cabin aft we were always below freezing point. The second winter, as opinions were divided, I tried letting the vessel lie without throwing snow over it. But although the winter was far milder, the temperature in the fore-cabin soon fell below freezing-point at night. When, thereupon, I had the vessel covered in with snow, the old state of things soon returned.

"URANIENBORG." THE ASTRONOMICAL OBSERVATORY IN SUMMER TIME.

"Uranienborg," the astronomical observatory, was the last of the series of buildings. One forenoon we all assembled to assist the astronomer in the erection of a building suitable for his purposes. Lieutenant Hansen, our astronomer, preferred the style of a rotunda, and we promptly set to work to build him an Eskimo hut. The structure was not a magnificent work of art, but it was built, at any rate. As a base for the instrument he used an empty barrel.

One morning, as we were standing on the forecastle, promoting the digestion of our breakfast with a chat and, as usual, keeping an eye on the hill-side for reindeer, one of us pointing towards the north, cries : " Here is more sport ! " Immediately preparations were made for the hunt. But Hansen remained standing by my side and seemed to be straining his uncommonly sharp eyes. " Well, Hansen, have you no mind to shoot reindeer, to-day?" " Ah, yes," he said softly, "but not *that* sort of reindeer, over there—they walk on two legs!" After this startling announcement I rushed down to fetch my field-glass, which I brought to bear on the "reindeer" flock, and, quite right, there were five men.

" Eskimo ! "

Now, we had been discussing Eskimo frequently and at great length, but for some reason or other we had all considered it most improbable that we should encounter any here. We were now near the end of October, and we thought the Eskimo were extinct, and had been relegated to oblivion. And here they were before us.

All the information we had gathered concerning these Arctic barbarians rushed back into our memories. We knew from old books of travel in these regions that the North American Eskimo were not always amicably disposed. But we had learnt from Ross and Klutschak that the Eskimo word " Teima " was the best greeting with which to approach them. It meant something like a right hearty "good-day," and we had rehearsed this

IVAYARRA. OGLULI ESKIMO.

word "Teima" in the most varied styles of pronunciation. However, we did not dream of being so foolish as to put our whole trust in one feeble word. The only right

course was to consider the newcomers in the light of
enemies, and our plan of campaign was laid. I was
going with two men to meet the enemy ; Hansen and
Lund volunteered. The rifles were carefully examined
and loaded to the utmost capacity of the magazine.
Down on the ice I drew up my troops and inspected them,
and even the most critical general could not have found
fault with their appearance and bearing. I myself threw
out my chest as well as I could, drew myself up, made a
regulation right-about turn, and gave the command
" Forward—march! " With my brave men close behind
me I advanced, casting a sidelong glance up to the deck
where the Lieutenant and the cook stood side by side.
It seemed to me that their expression, at the sight of our
little host, was not exactly one of admiration, not even
of seriousness. Well, I thought, it is easy enough to be
gay when standing well-sheltered on board, while we
were going forth to meet the uncertain, possibly death,
here on the open field.

The Eskimo were now at a distance of 500 yards and
were coming down the hillside towards our vessel. I
advanced in my best martial style, and behind me I
heard the tramp of my men in well-timed cadence.
When within about 200 yards distance the Eskimo
halted. Several strategic possibilities presented themselves to my mind, offensive, defensive, etc., but I
thought it safest to command a halt. My men bore
themselves splendidly, in faultless alignment, with their
feet set at an angle of 45°, and with a mien betokening

courage and confidence in their leader. I thereupon reconnoitred the opposing host. They appeared to be talking excitedly, pointing with their hands, laughing and gesticulating, without any noticeable indication of hostility. But suddenly they deployed in skirmishing order and advanced. Well, I thought, rather death with honour than saving our lives by craven flight. "Forward— march!" And on we marched, expecting every moment to see the enemy take their bows from their backs and level an arrow at us. But no! evidently they are of a different mind. Is this a ruse?

Suddenly, there flashed through my mind, heated with the excitement of warfare, the word "Teima!" and "Teima" I shout with all the power of my lusty lungs. The Eskimo stop short. But now our excitement can no longer be restrained, we must bring matters to a crisis, and we rush forward, ready for action. Then I hear the call: "Manik-tu-mi! Manik-tu-mi!" And this has quite a familiar sound — I well remember it from M'Clintock—it is the Eskimo's friendliest greeting. In a moment we fling away our rifles and hasten towards our friends, and with the universal shout of "Manik-tu-mi! Manik-tu-mi!" we embrace and pat each other, and it would be hard to say on which side the joy is greater.

Our friends greatly surprised me by their appearance. We had but recently left the ugly, flat-nosed Eskimo of the North West Coast of Greenland, and here we encountered a tribe of which some could be called really handsome. A couple of them looked like Indians, and might have served as the prototypes of some of the characters in Cooper's tales. They were also tall and muscular. In fraternal union, we all proceeded towards the vessel. I heard the click of the Lieutenant's camera again and again. By his side stood Lindström with his broadest grin. And I certainly can't say I felt much like a great general.

Our visitors accepted, with the utmost delight, our invitation to come on board. There were 100 reindeer carcases lying piled up on deck, and the Eskimo stared at this large store of meat, but said nothing. We stood for a long time talking to them, laughing and joking. Then Lindström asked me *sotto voce* whether we ought not to offer them refreshments. Yes! I told him to make coffee and put out some ships' biscuits. We took our visitors down to the hold; I did not care to show them into the cabin, as I was afraid they might leave visitors behind. The North Greenland Eskimo, at any rate, are notorious for parasites. Biscuits and coffee were served, but did not appear to be particularly to their taste. They indicated by signs that they would like to have something to drink, and when water was given to them their faces shone with delight. They drank a couple of pints each. But if they preferred icy cold water to coffee, so possibly they . . . .

"I say, Lindström, just hand me that old leg of meat lying there!"

Yes, I was right. That was something different to biscuits. Now we also discovered that they were not quite so unarmed as they looked. From the legs of their kamiks they drew large, long-bladed knives, and in no time they had polished off three joints, leaving only the bare bones.

Wiik and Ristvedt had not witnessed any of the scenes during the Eskimo's arrival, and as they did not come

FESTIVITIES AMONG THE NECHILLI ESKIMO.

down I thought they had no idea of what had been going on. When at last the Eskimo had finished their meal I signified to them to follow me, and led the way to Villa Magnet. There was no one to be seen outside, and I knocked at the door and went in. Ristvedt and Wiik were both sitting deeply engrossed in their books. The Eskimo kept behind me in silence.

The First Winter.

"What a strange thing!" I said, "that we should have visitors in these regions—and good friends, too! Allow me to introduce"——

Both gentlemen started, sprang up in a jiffy, and advanced with their most elegant bow; and the Eskimo

Ristvedt.   Wiik.
INTERIOR OF THE VILLA "MAGNET."

stepped forward. There was hearty laughter, in which the Eskimo joined with tremendous roars.

Of course, it was not very easy for us at first to make ourselves understood by these people. But when once we had got them to see that our desire was to learn what different things were called in their language, the *entente cordiale* soon made rapid progress. Certainly we

Chapter IV.

should scarcely have been able to keep up a regular ball-room conversation with their ladies, but we acquired a vocabulary sufficient for our requirements. And we had no balls there.

They stayed with us over night and returned to their homes next day. We had already managed to make them understand that we wished to buy dressed skins of them. The lieutenant's achievements and my own in the skin-dressing line excited their undisguised merriment, and we thought we had best keep them for ourselves. A couple of days later they returned, bringing with them some large fine reindeer skins. However, with a keen eye to business, they only brought us large skins of bucks, for which they themselves had not much use. But we paid them back in their own coin, and only gave them one sewing needle per skin. I now decided to accompany them on their way home to see where and how they lived. They had given us to understand that they were not going to take any sleep on their way home ; therefore it could not be so very far off. Next morning at 11.30 we started. I had with me a sledge on which I had my sleeping bag, a little food, and a quantity of things which I knew the Eskimo would value highly. Following the inborn promptings of civilisation I harnessed my visitors to the sledge while I myself accompanied them on ski, and away we went to the west at full gallop. The Eskimo did not use anything in the way of ski, snow shoes, or the like, as the hard snow, swept together into a compact mass by the storms, would bear their weight. And I had all my work cut out to keep up with them on my ski.

It was November 9th, and darkness set in early in the day. I therefore thought it would be necessary to hurry on. I did not know, at that time, that to an Eskimo it is a matter of complete indifference whether he travels in daylight or in the dark, in bright weather or in the thickest fog, in a storm or a calm, or in a snowstorm so thick that he cannot see his hand before him. I only learnt this later, on closer acquaintance with them. About 3.30 P.M. they made signs that we were nearing their camp. And from the top of a hill-crest I perceived some faint lights down in a snug, sheltered valley. By this time it was almost completely dark. The Eskimo uttered loud cries of joy, and were exuberantly happy at the sight. And indeed, the little lights down there seemed very alluring and inviting, suggestive of warmth and comfort, food and drink, and all that can cheer the wayfarer on an inclement and cold winter's night.

When we came within call, my conductors uttered loud shouts, of which I could only catch the one word : " Kabluna " (white man). And immediately the inhabitants of the camp swarmed towards us. It was, indeed, a strange scene ; I can still picture it, and shall never forget it. Out in the desolate snow landscape I was surrounded by a crowd of savages yelling and shouting one above the other, staring into my face, grabbing at my clothes, stroking and feeling me. The rays filtered through the ice-windows of the huts, out into the last faint dusky-green shimmer of fading daylight in the west.

Poetical reflections may be all very well, but not in a temperature of —4° F., and with an empty stomach. I was longing for a warm room and food, and accompanied Attira to his hut. I had taken a fancy to him. He and Tamoktuktu resided there with their families. It was a large hut, quite spacious enough for its eight inmates. Soon after our arrival, the male members of the colony assembled for a feast consisting of raw reindeer meat and water. Three entire carcases of reindeer disappeared before I could consume a sandwich. They were chatting and laughing all the time. But there could not be any efficient "Women's Rights'" Association here, as none of the women were present at the feast. When I tried to show them how we conducted ourselves towards our women, and courteously offered some meat to Mrs. Tamoktuktu, they shrieked with laughter, and evidently considered that I was a most irrational being. When the men had eaten their fill, the women were admitted. They greeted me with "manik-tu-mi," and stroked me nearly all over the body. Then they departed, without having been offered any food. I was subsequently set at ease on this point, being informed that even Eskimo women are not unmindful of their own bodily wants when left to themselves. The disappearance of a joint of reindeer causes no particular comment.

About 10 P.M. I went to bed in my sleeping-bag, which

The First Winter.

I had arranged on the common berth between the two families, and slept till broad daylight. But before then the Eskimo were astir. I saw them sitting upright, innocent of raiment, their sole covering when asleep being the skin rugs, enjoying their morning air-bath. A chilly pastime, I thought, and, snuggling down in my bag, I dosed off again.

WIIK AT THE ENTRANCE TO THE OBSERVATORY FOR THE MAGNETIC VARIATION INSTRUMENTS. GJÖAHAVN.

Their camp consisted of six huts and was situated near a large lake, which the Eskimo called Kaa-aak-ka. They told me, in fact, that it was just hereabouts that Lund and Hansen had been shooting in the autumn.

Later in the forenoon I returned home.

On November 2nd the "Fixed Station" started work. Wiik had fitted up the self-registering magnetic instruments in the "Variation House," and attended to these observations quite single handed. Every noon he changed the paper on the registering drum, and this was not always exactly a pleasure. Considering that he had to plough his way through wind and driving snow, often a yard deep, in a temperature of 76° below zero Fahr., one will readily understand that it needed an able and devoted man to carry out these duties. He did this for 19 months without intermission, a handsome monument, of his own creation, to his zeal and devotion. Meteorological observations were taken three times in 24 hours. Besides, we also had recording instruments, which were in action day and night all the time. Ristvedt was chief of this department. He had many a rough spell at his instruments, out in the cold, dark nights. All honour is due to him for his devoted and conscientious labours.

The office of astronomer to the "Gjöa" Expedition was a very arduous one. The expedition had had neither space nor funds to bring out a collapsible structure in which the observer could await in peace and comfort the right moment for astronomical observations, which in our case had to be carried out with a bare little snow-wall to ward off wind and snow, and in the very lowest of temperatures. Just imagine a temperature of 40° below zero and a biting snow. Till nightfall the sky has been hazy, but now it has cleared, revealing thousands of glittering stars. "Ah, what a beautiful star-lit night!" we others exclaim. But the poor astronomer must turn out of his cosy cabin, go behind his snow-wall, and suffer for hours all the peculiar hardships besetting the astronomer in the Polar regions—stiff-frozen fingers, an ice-coated telescope lens, and all kinds of discomforts.

Lund and Hansen were charged with all the work connected with the vessel. By way of extras, Lund had his "water-hole," and Hansen the dogs to attend to.

The dogs, unfortunately, suffered from the same disease that attacked them on the voyage out. First, Tiras, of the Godhavn team, was seized and died, and before Christmas we had lost seven of our best dogs. Too late, unhappily, I came to the conclusion that it was probably lack of fatty matter in their food that killed them. They had been fed all the winter on lean reindeer flesh.

Ah! those dogs! They have splendid appetites, as I know to my sorrow. They are always on the look out for some extra, wherewith to supplement their rations. The other day they helped themselves to an extra meal in a fashion as unexpected as it was unpleasant. Silla, who was in a highly interesting condition, had been shut up in the passage outside the "Magnet" to await her confinement. One fine morning, however, she managed to steal out, and proceeded straightway towards the vessel. Midway she was met by all her attendant cavaliers, wildly excited at seeing the lady again. They surrounded her and escorted her on her way. But it so happened that poor Silla was suddenly seized with labour, and her progeny had to content themselves with a snowdrift for a cradle. At a signal, given, of course, by Lurven, all the other dogs rushed at the pups, each snatching up one and consuming it on the spot. When Silla became aware that her pups had vanished, she raised herself and walked on. Again she was seized, and gave birth to her last puppy. Then, lest the other dogs should appropriate this one also, she hastily consumed it herself. This almost incredible scene is vouched for by eye-witnesses.

Lindström had charge of the kitchen. This word suggests a warm, cosy place, with well-scrubbed benches, floor strewn with pine branches, and walls covered with brightly polished cooking vessels. Alas, Lindström had none of these. It might be warm enough there, scorching hot, in fact, so that he had to rush out into the open air to cool himself. But when he went in in the morning to commence his daily task, everything was frozen. It was dark and weird ; the " Primus " stoves looked grim and cold ; pots and pans made his finger tips ache with cold as he touched them, while the rest of us still lay snoring in our warm bunks. No, it was not pleasant work, and Lindström deserves a high tribute of admiration for the meritorious manner in which he discharged his duties for three years with unfailing good humour.

The winter advanced quickly, and Christmas was drawing near. Our preparations for the festival were many and important, and, as everywhere else in this sinful world, they were chiefly in the direction of food and drink. The cook was hard at work early and late ; to him, certainly, Christmas could hardly have appeared in the light of a holiday. He baked and broiled, and we saw big dishes of cakes and pastry hurried out of our sight into Lindström's secret repositories. "They won't find them this time," says he, with his most cunning smile. But it is not so easy to baffle "them." It is the denizens of the "Magnet" of whom Lindström is suspicious and in constant fear. They are like ravens when cakes are about. In fact, they were always playing all sorts of pranks on the good cook. Though, candidly speaking, one can hardly blame them, as Lindström always walks straight into these traps; and when he discovers that they have outwitted him, he is himself the first to laugh heartily and enjoy the joke. Such men are invaluable as companions on a voyage like ours.

But it is not only the cook who has extra work at Christmas. We all have a little of it. We have to trim and tidy up, of course.

Our friends the Ogluli Eskimo had now visited us many times and in large parties. They used to come about midday, build snow huts for themselves, and stay with us for several days. As a rule, they departed at 8 or 8.30 A.M. as precisely as if they carried watches, but now that it is dark they have not even the sun to guess the time by. Towards Christmas these visits became less frequent.

At last Christmas Eve arrived, with brilliant weather. Each vied with the other in exhibiting a festive spirit; in gay dress and finery and clean-washed faces resplendent with soap, this was by no means a daily experience.

Chapter IV.

But the reader must not misinterpret me in the sense that we did not like washing. On the contrary, we were only too glad to get a wash, there was nothing we liked better ; but the cook had so little water that he could not spare any.

The day passed pleasantly and cosily, we ate and drank and sang, as people are wont to do on Christmas Eve. And, finally, there came the Christmas tree, which was the work of Lund and Lindström. An artificial fir-tree, covered with glazed paper ! Christmas candles were alight on it, and made the illusion of a real Christmas tree a very vivid one.

Then came the distribution of Christmas presents, undoubtedly the most important function of the evening. When we were lying off Fremnaes, our relatives, friends, and acquaintances had sent on board a large number of mysterious packages bearing the inscription " Not to be opened till Christmas Eve." Some had even had the foresight to put "Christmas, 1904, 1905, and 1906." And this was fortunate, as otherwise we should have undoubtedly exhausted the whole supply on this first Christmas Eve. The packages post-dated with such foresight were left unopened loyally, and not brought out before the proper time.

We distributed the presents by drawing lots, to prevent any jealousy, as even angels might not be proof against that. And fortune played very capricious pranks. Great hilarity was excited by some funny paper caps which were lying among a quantity of pretty articles of fancy

FIRST CHRISTMAS EVE ON BOARD THE "GJÖA," 1903.

needlework. Among the lots, Lindström drew a cap with the inscription " To the fattened Polar pigs!" We roared with laughter, and placed the hat on the head of the "pig." Our little phonograph was untiring in treating us to songs and recitations whenever we wished for them.

Outside, the scene was grand and tranquil. A most wonderful Aurora Borealis illuminated the whole sky. But the ever-shifting rays fill the spectator's mind with a feeling of unrest. It seems as though, on Christmas night, at least, they bring silent, flickering messages from the outer world—from home, where they are now celebrating *their* Christmas.

On Christmas Day we celebrated a double festival, as it was the 25th anniversary of Wiik's birthday. He was the youngest member of the Expedition, and one of the merriest, full of funny stories and anecdotes ; an invaluable entertainer on board. The big cakes which baker Hansen had presented to us on our departure were the most prominent item in the feast.

In the forenoon the old Eskimo Teraiu came on board, and was most hospitably received as a Christmas visitor. Teraiu was one of our oldest Eskimo friends, one of the five who figured in our first meeting. He was a man between 50 and 60, and a most jocular fellow. Among his fellow tribesmen he was held in little respect, notwithstanding his age ; they looked upon him as something next to an idiot, whom they barely tolerated among them. It will be seen later that he was not lacking in sense

Chapter IV.

But Teraiu did not appear to have come on any festive errand. His appearance and mien bore the stamp of deep depression, and tears stood in his eyes (the old humbug!). He jabbered and talked volubly, and that he was in distress about something was evident enough. But it was not easy to get at the cause. At last our united endeavours succeeded in unravelling the mystery. The remainder of the tribe had gone away and left old Teraiu and his family behind, in the most shameful manner, so that now he had nothing to look forward to but starvation for himself and family, unless we would take pity on them, and let them stay with us during the worst part of the winter. Of course, we were deeply moved by his pitiful tale, and told him that he might come to us with his wife and child. Moreover, I told him that I would go to Kaa-aak-ka and inquire into the matter as early as possible.

Teraiu and his family turned up very soon, and on January 2nd, 1904, we decided to go to Kaa-aak-ka. The weather was splendid, calm and clear. But the thermometer indicated −47° Fahr. Lieutenant Lund, Ristvedt, and I, with Teraiu and his family, all prepared for the journey. We harnessed eight dogs to the sleigh. We did not take much in the way of necessaries, as we only intended to be away for one night. Each man provided his own outfit, and I had left the matter of provisions to our worthy cook, who was an experienced Polar traveller. We others had little experience of these sledge trips.

The First Winter.

We went at a smart pace over the sound ice to the west, and after six hours' journey reached our destination. In the camp at Kaa-aak-ka everything looked very different from what it was when I was there last; it was empty and deserted, the snow huts looked more desolate, without people or any other sign of life. Teraiu's hut alone showed signs of being inhabited. Kayaggolo, Mrs. Teraiu, or, as we generally described her, "the old eagle," removed the snow block which had been carefully set up in front of the entrance to the hut, entered, and lit a fire. Teraiu himself went to the lake and cut a hole in the ice to get drinking water. With his miserable ice pricker it took him two good hours to make a hole.

Meanwhile we selected one of the best of the deserted huts and took possession for the night. The dogs rushed about, sniffing and ferreting for some addition to their day's rations—a pound of pemmican—but in vain. In this indescribably desolate place there was nothing to be found for either animal or man. In Teraiu's hut, Kayaggolo sat with her legs doubled up under her, singing her perpetual " Hang-a-ha-ya-ha-a !", and neither her person nor her song had any attraction for us. We therefore crept into our own hut. But that was not very inviting, either. In the first place an icy gust, at a temperature of − 58° Fahr., swept in through the open window hole in the roof ; and, secondly, the hut was full of thrown-away reindeer knuckle-bones ; wherever we put our feet we trod on them ; wherever we put our hands we touched them. This was both unpleasant and uncanny. But we would soon alter it. We got our things in and were going to settle down when we discovered that Lindström had forgotten to provide us with candles. I can't say that either the hut or our humour was greatly brightened by the small, ridiculous bean-sized tufts of moss, saturated with train oil old Teraiu had brought us by way of makeshift. We called them "light pastilles"; but, after all, anything was better than nothing. Lund who was cook for the night, after considerable difficulty, at last got the "Primus" stove into working trim, and the hot steam rose cheerily from the kettle.

At last we opened the provision chest. We were ooking forward in glad anticipation to what the good-hearted cook would have put in for us by way of an extra treat. One guessed "pudding," the other "cakes." A packet of chocolate—very nice. What, another! good —and a third—a fifth; eight big heavy packets of chocolate for *one* night! Well, that was only one side of the case. On the other side was the hard bread, and underneath it—Lund dived in, we stood by with our "farthing dips," in anxious expectation—hard bread, yes—more hard bread. That was all. Not even butter. Probably our amiable purveyor had considered that too much of a luxury.

We commenced to chop up the chocolate and throw it into the boiling water, to make the most of what we had. There was silence among us; but I should not be surprised if at this moment our good cook on board the "Gjöa" had a phosphorescent light playing about his head, which was not the halo of the beatified.

"Well, after all," said one of us, "it might be worse."

"Could it, indeed?"

"We *might* have been quite without food."

That was true, and our spirits rose again. The chocolate was ready, it was poured boiling hot into the cups. "Here's luck! . . . . whew!"

The brown liquid was no sooner in our mouths than it was out again; it was bitter cocoa. And no sugar— nothing of the kind, of course.

Our last consolation was that among the hard bread there was a spice cake. We might use that in place of sugar. The inventor of this plan was the first to try it, taking first a bite at the cake and then a sip at the hot cocoa, but whew! out again, against the nearest wall. "The beast has spilt the petroleum over the bread." The whole bread supply was soaked with paraffin oil.

I will not dwell on the state of mind in which we went to bed, with empty stomachs, and tried to sleep. We did not enjoy much sleep, for our teeth chattered with the bitter cold. Lund was up early and lighted the "Primus" stove. "We must get some warmth into our bodies, at any rate," he said, and again he put on the cocoa, left untouched last night. Bitter as it was, it was better than nothing. We were lying half asleep, enjoying the warmth of the "Primus" and anticipating with enjoyment what Lund was preparing for us. I was just about to doze off once more when I started up again on hearing him exclaim. " I believe the very devil is in it." And then he found that the "light pastilles" had dropped into the kettle, and the resultant train-oil cocoa, or cocoa and train-oil compound, was the most fearful concoction I ever tasted.

There was no more hope of any enjoyment here, and at 8 A.M. we all were ready to start on our homeward journey. The result of our investigations was that Teraiu and his family must be looked upon as being "deserving poor" and we gave them to understand that they might follow us back and stay with us for the winter.

The journey back was not effected at such a dashing pace as the journey out. Famished and sleepless as we were, we soon got tired, especially as we had had no training whatever. The weather was foggy and Teraiu led the way. At 1 P.M. we sighted the "Magnet" but the fog thickened immediately after. We went on and on, and the half hour which it ought to have taken us to get to the vessel, extended to an hour, an hour and a half, two hours. Teraiu now declared we were "paa vidotta" (quite close). We went on for some time in the thick fog. At last the moon came out and by her light we discovered that we had been wandering about without any definite direction. I proposed to Teraiu that we should build a snow hut and stay where we were till daylight. But he protested most emphatically that he would find his way through. So on we went in uncertainty, a state of things I always object to. But later on

The First Winter.

I took Teraiu to task sharply, and insisted on stopping.
He informed me with a self-satisfied air, that he had
"found the right way and that we should soon be at our
destination." And he was right; a few minutes later, to
our unspeakable joy, we saw a bright shining light, the
riding light, which those on board had hung out for our
benefit. It was 9 P.M. when we got on board, and we had
been travelling uninterruptedly for twelve hours. They
had heard our voices since about 1 P.M., and as we were so
long in coming they had hung out the light. However,
the fact of the matter is, that if we had not had Teraiu
with us, or rather, if we had not allowed him to guide us,
he being, of course, quite familiar with the country, we
should never have lost our way. From various circumstances it became evident that he had led us astray
intentionally, thinking he would get an extra reward for
bringing us back into the right way again.

When I subsequently met the villain, he always pre
faced his conversation with "Teraiu angatkukki angi"
("Teraiu is a great magician"). But I did not think fit
to reward him for his magic arts. The day after our
return he built his hut on the shore, and remained with
us till late in March. I was not sorry to have had him
and his family as boarders, as he and his wife were
useful in many ways. In spite of his age he was active
and had great stamina; he could run in front of a
sledge without much fatigue from morning till night.
He was peaceable and respectful in his demeanour, and
always in good humour and ready for fun. As a

builder of snow huts he was matchless, and of the greatest use to us. Kayaggolo, his wife, was about the same age as he was. Her face reminded one of an old shrivelled up winter apple. Take one of these, and cut two small slanting slits for eyes, a little jab for a nose, a rather larger one for a mouth, and you have Kayaggolo's image to a nicety. She was about five feet high, and so filthy that even the Eskimo jeered at her. Her son, Nutara, ten years old, was as filthy as his mother, but otherwise a very winning little fellow, very intelligent, and full of the quaintest tricks. His filthy state was chiefly the mother's fault. After a little association with us, the Eskimo generally followed our example in the matter of washing and keeping themselves tolerably clean. But the Teraiu family were practically no better in this respect when they left us than when they came.

When I returned in the forenoon from the Absolute Magnetic Observatory, I made a regular practice of calling on Teraiu, whose hut was on my way, for a little chat. During these visits, Kayaggolo sometimes treated me to some singing, the most appalling display imaginable. When she was carried away by her feelings, she went into a sort of frenzy, threw her head back, closed her eyes, and yelled her loudest. On these occasions I fled precipitately, but could hear her yells for some distance. Her husband and Nutara were mostly on board, sometimes forward, sometimes aft, and were universal favourites. What they liked best was to sit

outside the mess room and watch "Henrikki" preparing the food. And, although Lindström could not
bear Eskimo, his good nature often enough got the
better of him, and he would give them some tit-bit or
other. Teraiu and Nutara grew accustomed to our food.
But Kayaggolo held fast to raw meat and raw fish all the
time.

Nutara.   Lieut. Hansen.   Kayaggolo.   Teraiu.
A VISIT TO THE CABIN.

Christmas and New Year's day being now over, we
had to begin thinking in earnest of our long-planned
sledge expedition. The plans had been many and
various, as plans will always be in these regions. At
last we decided that I, with a companion, should try to
make my way to the magnetic station, and if all went

Chapter IV.

well, we would then push on with the mail to Leopold harbour on North Somerset. An auxiliary expedition, conducted by Lieutenant Hansen with one man, was to help us along as far as we found it expedient.

All sledges were brought out, examined, and repaired where necessary. Ski and snow shoes had to be seen to, the tents examined, etc. Workshops were established everywhere. Lund was putting the sledges in order; he had to prepare manageable provision chests, and do a good many other things. Hansen, who was very skilful, was asked to achieve the most incredible tasks. Whenever any work requiring particular neatness and precision was to be done, Hansen was the man for it. He was also an expert in the use of the sewing machine. Ristvedt had his smithy down by the provision shed and his engineering workshop in the "Magnet," and did wonders at everything. Wiik did delicate mechanical work as repairer of a number of instruments, and Lieutenant Hansen devoted himself to scientific studies and displayed great energy as a glove maker. His skill in repairing old knitted gloves was unique.

As in the case of all sledging expeditions on the Polar ice, the question of sleeping bags was very hotly discussed. Our little trip we made with Teraiu to Kaa-aak-ka had convinced us that great improvements were needed in this direction, and there was a regular competition among us for the honour of devising the best patent. In the first place the sleeping bag was too wide, and had to be taken in to a considerable extent. A sleeping bag ought to be just wide enough to come close up to the body all round, yet, of course, not tight enough to compress. If it be so large that you have to curl up in order to touch the sides, you will never be warm. The patent bag with an entrance through a hole at the top and a cord for drawing it in close round the neck was the one most generally favoured. For my part I preferred it to all the other designs, and recommend it to everybody.

Our tents were made up like Eskimo tents and were excellent. They required no repairing. They could be put up by one man, single-handed, even in the strongest wind, and when once properly fixed up they were never blown down. They stood the test on more than one occasion. However, we made one improvement in them, and that was in the door. The doorway is always the awkward point in a tent, especially in the Polar regions. Very frequently the patent closing arrangements consist in a number of fastenings and lashings, so that after having got inside you have a great deal of trouble to close the door after you. However, I have never seen any tent door that will really keep out the drifting snow in a snowstorm, except the one we made ourselves. As two of the members of the Expedition hit on almost exactly the same idea I will not mention any name as that of the inventor. The patent was so simple that we have had many a good laugh over it. But the most ingenious is frequently the simplest. We simply sewed the mouth of a sack round the entrance of the tent; we then cut a hole in the bottom of the sack, and through this hole we went in and out, afterwards tying the sack up with a cord. A better tent door does not exist. It is easy to open and close, and is absolutely secure. The many-sided problem was thus solved—by means of a sack.

Experience taught us that tents were too cold in a temperature below −22° Fahr., and we therefore decided to build snow huts, as from what we had seen among the Eskimo these were much warmer. It certainly takes longer to build a snow hut than to put up a tent, yet I consider that a good comfortable night's rest after a day's toil is so important that I would willingly expend the extra hour's labour required to insure it. After a bad night's rest one is not fit for work next day.

We therefore set about studying the art of snow-hut building. We had no lack of snow, nor of time, and in Teraiu we had an excellent tutor. At first we let the old man do all the work while we carefully watched his methods. We soon saw that one essential condition is to procure snow having certain peculiar qualities. But this requires much experience, in fact, almost an inborn talent. The Eskimo uses a simple appliance, called the "hervond"—a reindeer-horn staff of about 40 inches long, and as thick as a stout walking stick. On one end it has a handle of reindeer bone, and on the other a ferrule of musk-ox bone. The

The First Winter.

Eskimo are also gifted with a wonderful instinct for finding just the right place where suitable snow is likely to be found. If they have not their "hervond" with them they use a long-handled knife which they always carry in a strap on their backs when they travel. We never acquired any degree of perfection in finding the right sort of snow, but we succeeded to some extent.

Armed with a huge knife, we four, the Lieutenant, Ristvedt, Hansen, and I, met every morning after breakfast outside Teraiu's hut, to call him. If we came as early as 8 A.M., the family were invariably still in bed. Teraiu jumped up and dressed himself in a jiffy; Eskimo clothes are large and ample, and are easily slipped on or off. It took him longest to put on his foot-gear. The Eskimo is careful of his feet, not only in dread of getting them frozen, but also for fear of sore-footedness, as he moves about on ice and stone-hard snow all day. He does not content himself with less than a five-fold foot-gear. When Teraiu had at last put on his outer "anorak" (coat) he wore no undercoat on these short outings—he came out. We each had our special day. Hansen is the most talented of us all in the architectural branch; his huts were always real masterpieces. We soon found a site with a good and plentiful supply of snow in the many small valleys which lead out of the harbour. It took us, as a rule, an hour and a-half to erect a hut large enough for all four. When the work was finished, we assembled inside to criticise its merits. Teraiu was always quite as enthusiastic as we were. "Mamakpo, mamakpo!"

Chapter IV.

(excellent) he would exclaim. The special object of his joy, however, was the anticipated reward. He did nothing without remuneration. However, it was not a large matter, just a piece of wood or iron, or whatever it happened to be, for at that time he was collecting materials for making a sledge. His demands were not exorbitant; he was fully satisfied with a three-foot plank. " His family and their paraphernalia were not large, and therefore the sleigh need not be a large one," he opined, with a philosophy which many might imitate.

During these days, Wiik and I simultaneously made observations in order to determine the most important question, whether our instruments were in perfect order. Wiik did so in the Fixed Absolute Observatory, while I, for my part, had built a new observatory of my own, 250 feet away. With all these buildings, there was something like a little town lying around Gjöahavn. The results of our tests were quite satisfactory, and as far as the instruments were concerned, we could start with confidence. For geographical observations I had brought out with me a very small theodolite Nansen had lent me. He had had it with him in Greenland. According to the astronomer's very careful examination, this was also in order.

Now there remained the important work of packing our sledges. It was necessary to keep them under cover during the packing operations. We found it impossible to build a hut to cover the two big sledges, However, we applied to Teraiu, and to our request he grinned most significantly, and stretching forth both arms, his eyes glistening with greed, he said : " Panna angi," " Big knife." He wanted a big knife for building such a giant hut ; and it was promised him. We started at once on the building. For the sake of convenience, Teraiu designed it in an oblong form, and in order to have the sledges near, the hut was built on the ice close to the vessel. The hut assumed really gigantic dimensions, and quite a scaffolding had to be raised for Teraiu when he was about to construct the roof. With our help he completed the colossal hut in four hours. This was on a Saturday, and we had to suspend work and let the hut settle down till Monday. We were inexperienced at that time, and did not know that the hut ought to be heated inside in order to consolidate it. Teraiu, the sly fox, said nothing. His idea, of course, was that as the hut collapsed he would earn another big knife for building a new hut. But this time he had miscalculated. The hut collapsed, indeed, on the following day, but he had to build a new one without any further recompense ; we told him that was our custom. The new hut was built a little way up the beach because Teraiu feared, perhaps rightly so, that the movement of the ice had caused the collapse of the first hut. The new hut proved sufficiently strong, and our two sledges were put into it and the loading commenced. One sledge was to carry a load of 7 cwt., and was to be driven by Hansen with our seven dogs. The second was to carry a load of $5\frac{1}{4}$ cwt., and was to be drawn by us three.

Chapter IV.

On February 28th we put the last finishing touches to the work, and on the morning of the 29th we all brought the sledges up to the ridge so as to have them in a convenient position, ready for starting. Up on the crest of the ridge we built a high snow wall all round the sledges, and placed large snow blocks upon them, to keep the foxes off. Then we returned and spent the last night together on board.

I looked forward to the expedition with confidence. We were well equipped, and had good reliable comrades, and smart dogs. We should have been glad of a few more of the latter. But with a good heart and a good will we would manage with those we had.

Amundsen.   Helmer Hansen.
Lieut. Hansen.

THE FIRST SLEDGE EXPEDITION.

CHAPTER V.

Towards the Pole.

On March 1st we were ready to start. The thermometer stood at 63½° below zero (Fahr.). But in the course of the month of February we had become so used to cold that it really did not make any great impression on us. We were, indeed, extremely well clad, some of us in complete Eskimo costume, others in a partly-civilised style. My experience is that the Eskimo dress in winter in these regions is far superior to our European clothes. But one must either wear it all or not at all ; any mixture is bad. Woollen underclothing absorbs all the perspiration and soon becomes wet through and through. Dressed in nothing but reindeer skin, like the Eskimo, and with garments so loose and roomy on the body that the air can circulate between them, one can generally keep his things dry. Even if you are working so hard that you can't help getting wet, the skin dries again much easier than wool. Besides, woollen things soon become dirty and then they do not impart much warmth. Skin clothing keeps nearly as well without washing. A further great advantage of skin is that you feel warm and comfortable the moment you put it on. In woollen things you have to jump and dance about like a madman before you can get warm. Finally, skins are absolutely wind-proof, which, of course, is a very important point.

Our comrades who were to remain behind followed us up on to the ridge as far as the sledges. The dogs were put to, a last farewell, and then we were off. Hansen drove the dogs for one sledge, but got into harness himself as well. All the seven dogs were young animals and they found it hard to manage the load. Lieutenant Hansen, Ristvedt and I were harnessed to the other sledge. The ground sloped so slightly upwards that it was barely perceptible, yet it made itself felt all the same. The first hour, when all were fresh at it, things went very well. But then the difficulty began. Hansen managed the dogs well. When he noticed they were going to give in, he took a turn in the traces so that they, feeling they had some additional help, pulled away again. But we three were not so well off with the other sledge. It seemed as though we were driving it through the sand of the desert. Even at home in Norway we know how tiring drift-snow can be, but in the severe cold up here it is much worse. Every now and then the sledge stopped ; every little snow-drift meant a stoppage. One —two—three—helloa ! and over it we went, then on again, but not for long. A fresh drift, a fresh stoppage, and another tug.

About three in the afternoon we determined to pitch our camp. It had begun to grow dusk, and before we put up our snow hut it would be dark. Now we had to find good snow. We were out in the middle of a great lake, and the snow was not good anywhere ; we hacked away with our knives, but it was too shallow, The shore was too far away, we should not reach it till after dusk, so there was nothing to do but to remain where we were.

First we let the dogs loose. They had had hard work of it, and deserved their freedom and rest. The boss among them was Fiks, an unusually fine greyish white dog who had risen to be lord of the rest simply by his commanding character, and not at all by reason of his strength. If it came to a fight Fiks would get a good drubbing. But he seemed born to rule, and was obeyed. Syl was his grand vizier, the ugliest dog in the whole team : dark brown and with a stupid suspicious expression. The pointed erect ears which give the Arctic dog such a wideawake look, in Syl, stuck straight out and made him look more stupid still. The instant the harness was off, Fiks made his round, followed by Syl, from dog to dog. And in sign of subjection everyone had to lie on his back with all four paws in the air before his highness Fiks. If one delayed doing so, Syl was at him like a flash. Syl got his name from his sharp-pointed teeth. So the dogs had their food, and we were left in peace to begin our building operations.

We put on our building gloves, made specially for constructing snow huts. They had long cuffs, to prevent the snow from penetrating the armholes, and had to be securely tied. We set to work, each armed with his half-yard snow knife. The architect appointed for the

Chapter V.

occasion marked out the ground plot in a circle, and along this line he kicked out a furrow four inches deep to afford support to the foundation blocks. We cut out the blocks, and he attended to the construction. An igloo—as the Eskimo call their snow huts—is erected in a spiral form, something like a beehive, and always against the sun, *i.e.*, from right to left. The blocks must

be two feet long by one and a-half feet high and four inches thick. The greatest difficulty consists in working the wall inwards to form the roof. Any duffer can put up a straight wall. As the thermometer showed $70\frac{2}{3}°$ below zero (Fahr.), no one was tempted to be lazy, and the work went rapidly forward. The cook appointed for that evening set to work inside the walls as early as

possible to get the building warmed up quickly, as well
as to get the food ready. The workers outside re-
doubled their efforts as they began to scent the odour
of the food. The last task was to go over all the chinks
through which the light shone from the inside and stop
them up. Then we inspected the sledges to see that

MEMBERS OF THE EXPEDITION, BACK VIEW. (1903, 1904).

all was well lashed and covered up, especially from the
dogs, who were great thieves. Poor beasts, they coiled
themselves up as well as they could to leeward of the
hut walls and sledges, warming their noses under their
tails. The hut was ready, a last look out into the great
stillness, with the western sky in a dying glow of green
and the stars growing in brightness ; then we knocked

the snow off our clothes and crept in. And this I will say, that happier people could hardly be found on the earth than we four fellows in that sheltered, cosy hut round the steaming hot food, only separated by a wall from the waste and the biting frost without. After the meal out came our pipes, and it was only the thought that we must rise early for new toil next morning that made us break off our chat and creep into our sleeping bags. The day's efforts soon made themselves felt, and four mens' regular breathing suggested that Morpheus, the friend of mankind, may possibly also have been an Arctic traveller.

At five o'clock we were awakened by the rhythmical pump strokes of the Primus (the stove). The cook had not overslept himself. It is strange that everything seems so much less attractive in the morning than in the evening ; thus, for example, our comfortable hut now looked gloomy and narrow. A cup of steaming chocolate, however, improved conditions considerably. One of us complained that the boots he had used for a pillow might have been a little softer. His consolation was only a desultory question from another why he did not keep them on, then they would not have got frozen stiff.

I was very anxious to see what was the reading of the minimum thermometer I had set outside in the evening. The temperature had fallen the day before so suddenly— from — 65° to — 70° Fahr.—that I thought it had dropped still further in the course of the night. I removed the snow block which closed the entrance and crept out. There was a feeble ray of daylight and a dead calm. The stars seemed unusually bright and large, indicating intense cold. I cannot say that I felt it myself. But the night's minimum was — 79° Fahr., a pretty sharp frost. We could not but praise our excellent outfit, which together with our good snow hut had kept out this cold. God knows, we felt it in our finger tips when we had to take off our gloves to work if we found them in the way. Our fingers turned white in an instant, and we had to get life back into them sharp, either by putting the gloves on again, or by clapping our hands together, or, better still, by adopting the Eskimo dodge of putting them next to the bare abdomen.

The dogs lay as we left them in the evening, with the exception of Fiks and Syl, who, of course, had got loose to make a row. The problem of securely tying up the dogs was one we never solved. Somehow or other they always broke loose, if they were so inclined. Some would keep still, but when one got loose there was no end of a row, accompanied by baying and envious howls from the others. Little Bay went by the name of Ola Höiland, because no chain could hold him, and Lilli played the trick of puffing her neck out when we put her collar on and freeing herself later on. It was no joke to have to turn out of one's sleeping bag in the middle of the night and go out to keep the dogs in order.

On we toiled again. After yesterday's experience, we put the wooden runners on under the German silver mountings, as the sledges ran far better on wood in the sharp cold. One can't do better in these matters than copy the Eskimo, and let the runners get a fine covering of ice.; then they slide like butter. But we had not had any experience of this. The speed indicator was applied to the dog sledge; it was an old apparatus from the second "Fram" Expedition, but in excellent condition. In spite of all our efforts we progressed so slowly that the wheel seemed to stand still. What added to our trouble was a sharp head wind, very biting to the exposed parts of the face. We were continually watching each other's faces, and detecting a white nose on one or a frozen cheek on another. We did as the Eskimo do; we drew our warm hands out of the gloves and applied them to the frozen spot till the blood came again into circulation. I had long given up the old household remedy of rubbing with snow; it was not known among the Eskimo. While the wretched little wind and the 58° below zero struck us like needles or whip-lashes, the dogs did not seem to mind; but the poor creatures suffered greatly, especially in the early hours of the morning, when they were still tired and stiff from the previous day. We, too, had a rough time of it. I now saw there was little to be gained by going on in this way. As no change occurred in the temperature either on the second or third day, after consulting my companions, I decided to turn back and wait for milder weather. Early on the third day we brought part of our things into the igloo, as a depôt, and walled it up again carefully. The situation of the spot was accurately determined, a flag set up on the roof, and a photograph taken ; and then we shaped our course back to Gjöahavn. The dogs soon saw which way we were going, and we men were all glad we had given up our hopeless task. The result was that we did the journey of seven miles in four hours, though in coming out it had taken us two and a-half days. But our sledge load was now considerably lighter. At 11 o'clock in the forenoon we surprised our companions on the " Gjöa " by our unexpected and hasty return.

The time was now occupied with work of various kinds. We had not been long away, but our experience was valuable, and we effected many alterations in our outfit as the result of it.

About this time some of the dogs got tapeworm. We had no medicine for this disease, but Ristvedt, who, independent of his other excellent qualifications, was a veterinary surgeon, managed the worms.

The effects of the advancing sun, which rose perceptibly higher every day, began to show themselves. Large bright surfaces appeared on the snow ; the intense cold had subsided, and sledging was considerably easier. Little by little signs of animal life reappeared. On March 12th we saw the first ptarmigan. One fine day Teraiu cut a hole in the ice, erected a snow wall as a shelter against the wind, and started fishing. The result was not great, but he succeeded in catching a dozen small cod.

Chapter V.

On March 16th I resolved to go out again and try to carry the depôt a little further. I chose Hansen to accompany me. The idea of a relief expedition was abandoned. Lieutenant Hansen's time would now be occupied in charting the station and erecting cairns for the purpose ; this was a laborious as well as a protracted operation.

Our second start was much more favourable than the first. It was fine weather, and the temperature— —40° Fahr.—was fairly seasonable. We took with us from the vessel one sledge with ten dogs. The effect of the sun on the snow was immediately noticeable. For long stretches the sledge went at a furious rate over the most brilliant snow-crust, so that we had difficulty in following. In about three hours we now accomplished the whole of our former toilsome journey. The igloo, with its depôt, was in very good order, and we at once began to divide the load between our two sledges ; there was an additional sledge in the igloo. The weight on each sledge was about 4 cwt., and to each we harnessed five dogs. Allured by the many signs of spring I had taken a tent with me this time. We were now only two men, and the building of a snow hut would take us a long time. Meanwhile we slept that first night very snugly in our good old igloo. Early the next morning we were on the road. In the calm weather we went swimmingly over the plains of King William Land. We were soon down in La Trobe Bay, on the east side. We skimmed smoothly over the even ice in the bay, and after dusk we erected our tent under a hummock. But we had bitter experience of the difference between a tent and a snow hut. We could not get warm even in the sleeping bags, and we passed most of the time turning and twisting about and knocking our feet together. It was pure enjoyment to be on the move again the next day, and get warmth into our bodies by means of a little hard marching. There was a frosty mist, which was bitterly cold. Unfortunately, we had spoilt our thermometer, and could not determine the degree of cold. Our petroleum, however, acted as a sort of thermometer. When it was thick and milky white, we knew it was about 58° below zero. We set our course north to reach Matty Island. I had proposed establishing the depôt on Cape Christian Frederik, where we had been on the sea trip southwards.

The sun peeped out from time to time, so we determined our position. Otherwise we took our bearings from hummock to hummock. Between these the ice, as a rule, was smooth and bright. The hummocks were not large, and were formed by newly frozen ice, as could be seen by the thin pieces of which they were formed. We halted now and then to make sure of the direction, as well as to rest a little and have a chat. At 10 o'clock we stopped to lash the sledges tighter, and the conversation turned on the Eskimo, whom M'Clintock met here in 1859; should we find some tribes here still? As we sat, we saw a black dot far out in the ice. Hansen, with his excellent sight, soon concluded it was an

Chapter V.

Eskimo approaching us. Shortly after several of them emerged from the hummocks, and very soon we could see thirty-four men and boys at a distance of about 200 yards. They stood still and observed us without any sign of coming nearer. I now felt considerably safer than on my first encounter with Eskimo; my acqaintance with the language was also better, and I decided to go to them. However, we got our rifles ready, and Hansen kept an eye on them. When we were quite near, I called out " Manik-tu-mi ! " and it was as if an electric shock had gone through the whole crowd. A 34-fold " Manik-tu-mi " was heard in reply, and I went straight up to them. Hansen, who saw that there was nothing to fear, abandoned his post and followed me. The Eskimo's delight, nay, enthusiasm, was really touching. They stroked and patted us, laughed and shouted " Manik-tu-mi " unceasingly. They were Nechilli Eskimo. They told us they were on the way to their seal fisheries; each man carried his spear in his hand and had a dog following him on a leash. They were also provided with large snow knives. They gave us the impression of being cleaner and better clad than our first friends, the Ogluli Eskimo. When I asked where their camp was, they pointed eastwards beyond the hummocks. I was anxious to make the acquaintance of these people, and told them that I should be happy to accompany them to their camp. They were frantically glad to hear this, and at once set to work to help us with our sledges, harnessing all their own dogs to them. We

had plenty of dogs now. When we had finished putting
them to, an old fellow came driving along on a little
sledge. This was Kagoptinner, *i.e.*, the "grey-haired,"

KAGOPTINNER IN HIS SON POIETA'S HUT.

who we later got to know was the oldest and best
medicine man of the tribe ; after a friendly greeting, his
three dogs were added to our team, and the old chap
himself was set on the top of one of our sledges. As

Chapter V.

we started off at a rattling pace, old Kagoptinner had his work cut out to hold on; the sledge was more often on one runner than on two. Some of the youngest boys capered about at the head of the procession, accompanied by the dogs, in the wildest confusion. They were beyond control. The Eskimo dogs were overjoyed at returning home so early, and our own got scent of the camp, and made for it. Then one of them suddenly flew at his neighbour, and before long the whole of one team was engaged in a furious fray. This was more than the other team could stand, and first one, then two, three, and at last the whole lot were engaged in mortal combat. The Eskimo threw themselves among them; snorting dogs and howling Eskimo formed one chaos, until they finally succeded in getting the dogs parted; the traces were disentangled again, and the journey continued. The men formed a row alongside each team, running, laughing, and shouting unceasingly. They were clumsy and heavy, but looked as if they could keep the pace a long time. In about an hour they began to shout "Igloo! Igloo!"—and sure enough, far ahead among the hummocks, we sighted a crowd of huts shaped like hayricks. Another half-hour and we were there. This was the largest camp I had seen, sixteen huts altogether. They were not arranged on any kind of system, but spread about according to the conditions of the snow.

The whole place looked quite deserted. We halted a little way off and loosed the dogs. The men made quietly for their huts, and shortly afterwards the fair sex made their appearance. They arranged themselves in single file one behind the other. When all were mustered, the strange procession started running towards us. At the head came old Auva; after her, her friend Anana. "Running" hardly expresses the movement: they reminded one of a row of waddling geese. They made straight for us, and I trembled; would they kiss us as a sign of welcome? Old Auva was appalling to behold. We had come upon them so abruptly that they had had no time to complete their toilet. Such clothing as Auva had on, was covered with fat and soot, her face shone with train oil, and her greyish black hair hung in wild confusion under the hood that had slipped down at the back of her neck. I looked at her with horror as she came nearer and hid myself hastily behind the little-suspecting Hansen, to let him take the first shock. Nor was Anana beautiful, either; she was covered with dirt and soot, and train oil, but anyone who could survive Auva, could easily put up with the other. Now they were up to poor Hansen, and I was just expecting the kissing and embracing to commence, when they swerved aside and formed a circle round us, emitting all sorts of weird grunts, and then waddled off back to the camp. Now that the fright was over, I was able to examine them more calmly, and I must say that my first impressions of Nechilli Eskimo ladies did not redound to their advantage. Whether it was pure accident that just the ugliest of them came to us then, or that my taste altered later, I cannot say; certain it is that I afterwards thought

Chapter V.

some of them were quite good looking. When the procession, which, in fact, had been organised in our honour, and to show us a welcome, reached the camp, those who had taken part in it retired to their respective huts.

Now we had to think about getting our snow-hut built. After the experiences of last night, we were not anxious to try the tent again. We selected a spot not far from the others, and set to work. At first, the Eskimo followed us with inquisitive glances. They, no doubt, hardly thought that a "Kabluna" (foreigner) could manage a piece of work which was their own speciality. But they did not wait long before very audibly expressing their views on the point. Hansen and I did something or other they were not used to, and in a trice the whole crowd burst out into noisy exultation. Their laughter was uncontrollable; the tears ran down their cheeks, they writhed with laughter, gasped for breath, and positively shrieked. At last they recovered sufficiently to be able to offer us their assistance. They took the whole work in hand, but had to stop every now and then to have another laugh at the thought of our stupidity. In a short time, however, the most beautiful igloo was ready for us. We took our things in and arranged them, and then we went round visiting.

I had already noticed one man among the rest out on the ice. He was not like his companions, full of laughter and nonsense, but rather serious. There was also something haughty in his air, almost commanding, yet he

NECHILLI ESKIMO IN THEIR SNOW HUT.

could hardly be a chief of any kind, as the others treated him quite as an equal. A fine fellow he was, with raven black hair, and unlike his fellow tribesmen, had a luxuriant growth of beard ; he was broad shouldered and somewhat inclined to corpulency. His belongings— clothes, tackle, dogs, etc., were choice in quality and appearance. When I came out of my hut, he stood at a little distance from the others and regarded me with a look that seemed to intimate that he had something special to tell me. I accordingly went straight up to him, and he bade me go with him to his hut. It looked exceptionally neat outside. Like a courteous host he made me enter first. This, as I am now inclined to think, was an accident, but at the moment it increased my sympathy for the man, as was only right and proper. His name was Atikleura. He was a son of old Kagoptinner, the medicine man we had met on the ice, in his own turn-out. He showed himself later to be far superior to all his countrymen in every respect. I followed his suggestion, and went inside his igloo. A passage led into the hut proper ; this was so low, that I had to stoop down. It had two extensions, like quite small huts, and what they served for was not difficult to guess by the odour ; there was nothing to see, as the dogs were the scavengers. A hole so small that one had almost to creep through it led into the dwelling room. When I stood upright inside, I was speechless with astonishment. It was quite an apartment for festive occasions ; it had been constructed the day before, and was therefore

Chapter V.

still gleaming white. From floor to roof the room measured fully twice a man's height. The blocks in the wall were regular and of equal size, and the inside diameter was not less than fifteen feet. It was evident that Alikleura knew how to build beautifully. The sleeping shelf was so high, one had to swing oneself up on to it, and it was covered with the most delicate reindeer skins. Everything gave the impression of the most perfect order.

On the form before the fire place sat the lady of the house. She was strongly Mongolian in type, and by no means beautiful. But she looked clean and tidy. Like most other Eskimo women she had lovely shining white teeth and beautiful eyes, brown on a light blue ground. She was tattooed like the rest on the chin, cheeks, brow, and hands. We learnt afterwards that these women also tattooed themselves on other parts of the body. Her manner was not so engaging as her husband's; on the contrary it was somewhat brusque. Her three sons had also evidently much respect for their mother. The eldest, Errera, was a youth of sixteen or seventeen, of the purest Indian type. The absolute dissimilarity between the child and the parents was then inexplicable to me, but it became less so later when I learnt to know their matrimonial relations better. Errera was a very sympathetic, one might almost be tempted to say, well-bred boy, whose polite and pleasing disposition endeared him to us all. The next in age was his exact opposite, a saucy fellow, who had been given as a present to grandfather Kagoptinner, who, in grandfatherly fashion, spoiled him and withdrew him from his mother's good influence. The youngest was Anni, a perfectly charming little chap of five, his parents' darling. The whole family was better clad than the other Eskimo. The boys' clothing was of quite a model type. From what I saw I determined to be on good terms with Atikleura. He was manifestly a man it would be an advantage to know.

As soon as I came in, Atikleura fetched a skin sack, out of which he took a very finely-made reindeer skin garment which he presented to me. In my eagerness I wanted to strike while the iron was hot, and hinted that I should greatly value a suit of underclothing as well. Evidently very pleased at my request he now brought out some old, worn underclothing, put them on in place of those he was wearing, and handed me the latter with every indication that I was to change there and then. Somewhat surprised, I hesitated; I must say I was not in the habit of exchanging underwear with other people, especially in the presence of a lady. But as Atikleura insisted, and his wife, Nalungia, showed the most complete indifference as to what I did, I quickly made my decision, seated myself on the form, veiled my charms as well as I could with the bed clothes, and was soon clad in Atikleura's still warm underclothing. After this I was regaled with water, frozen raw reindeer meat, and salmon, served with small squares of seal blubber. I did not relish the meat, but the frozen

Chapter V.

salmon was quite delicious in flavour. For dessert I had frozen reindeer marrow, which did not taste badly. Atikleura also provided for our dogs, and dealt out huge lumps of blubber to them. This unwonted fare vanished like dew in the sunshine.

After this feast of welcome was over I put on my fine new outer clothing and went out. Outside in front of the hut lay a very fine polar bear skin; thick-haired and shining white, a really splendid specimen. I stood gazing at it in admiration, but then went hastily over to our own hut to bring some return gifts for my friends. Luckily I had brought with me some sewing needles, spear points, etc., on which the Eskimo set special value, and I think Atikleura and Nalungia had hardly ever been so happy in their lives, as when I brought them my gifts—two spear-points for him, and six sewing needles for her. After this I made a round of all the huts, and was everywhere very well received. Old Auva was particularly amiable. When I took leave of her, she presented me with a little bear skin and two reindeer tongues. As the latter were thoroughly filthy and covered with hair, she first picked the coarsest dirt off and then had recourse to the universal Eskimo tool—her tongue. She licked my reindeer tongues so clean that you could see your face in them. On my return to our own hut, Atikleura stood there with his bear skin. He handed it to me, beaming with joy. As a modest young man, I represented to him that I could not possibly accept such great generosity. But Atikleura would not hear of it, and resolutely carried the skin into my hut and laid it there.

Hansen now returned. He had also passed the time in paying visits and taking 5 o'clock tea in the different huts. Like me, he had profited by the opportunity to rig himself out, and was enraptured with all he had seen and heard. As a present he had received reindeer tongues, which had evidently been treated in the same manner as mine. We decided, however, to brown them a little more before eating them. Our arrangements with the stove and other cooking apparatus interested the Eskimo in the highest degree, so that the hut was soon full of visitors. The women kept away, probably by order of the strict husbands. Only Auva and Anana, who were both merry widows, ventured to pay us a visit. We formed quite a high opinion of the morals and manners among the Nechilli Eskimo. The men seemed to watch over their wives, and the latter to be faithful and obedient to their husbands. This good impression, however, did not last long.

We had decided to travel further northward next morning, and accordingly made ourselves ready the night before. As our dogs were rather tired, I applied to Atikleura to see if it was possible to get the loan of some dogs from the Eskimo. In Ogluli Eskimo dog is called "miki," but when I asked Atikleura for "miki," he was a long time before he understood me. I explained over and over again what I meant, and at last he seemed to understand. He nodded assent, and I was satisfied.

Chapter V.

A younger brother of Atikleura, Poieta, had promised to accompany us north to show us the best way through the hummocks. He was a fine fellow of twenty-five or twenty-six, not so heavily built as his brother, but with an open, engaging face. He was smart and willing. His wife, who also bore the name of Nalungia, received some sewing needles for her husband's services.

The first who met me, when I came out next morning, was Anni, Atikleura's youngest boy, and the apple of his eye. He stood, evidently waiting for me, and smiled with a mild and pleased expression. I took the boy by the hand, and went with him to his parent's hut. Here Atikleura was already at work on a piece of bone, while he hummed and sang. I greeted him, and told him to be good enough to get his dogs ready, as we wanted to be off. Astonished, and a little impatient, Atikleura pointed to the youngest and said : " Ona mikaga ! there is my boy ! " " Miki," in Nechilli, means child, not dog. The misunderstanding was soon cleared up, and we got the loan of two good dogs, while Atikleura and Nalungia retained their darling.

When all was ready and the dogs harnessed, I had all the women of the camp called together. They were arranged in a row and passed me one at a time, while each of them got four sewing needles as thanks for good treatment. That the notorious feminine cunning is not an exclusive possession of white women, was shown by old Todloli, who, when she had received her needles, sneaked in again at the queue and came up for a new supply. When she saw she was found out she burst out laughing heartily, all the rest joining in. Taking them altogether they were the merriest people I have met.

We started off smartly, Poieta leading. He knew the way and kept us clear of hummocks. At 4 o'clock in the afternoon we came into high pack ice, and Poieta halted. As it cleared for a moment we caught sight of Matty Island. It was a sheer delight to build an igloo, when one had Eskimo help, and the whole was done in an hour.

The next day was miserable. The wind blew right in our teeth, and judging by the state of the petroleum, the temperature was about 58° below zero. Time after time I got my nose frozen white, and big chilblains formed on my wrists. Hansen managed better. His nose suffered a little, but his gloves closed better over the cuffs of his coat, and protected his wrists from the frost. At noon we hit upon a little Eskimo camp of six huts. And now Poieta refused in the most decisive manner to go further, for which we could not blame him, as the weather was so abominable. These Eskimo were on an average taller than the Nechilli, and stood about six feet high. But otherwise they produced a much less favourable impression. They had the failing of begging for all they saw. So troublesome did they become with this during the evening, that we thought proper to creep into our hut and shut ourselves in. They had, of course, helped us very kindly in erecting our igloo, but I had no trust in them, and before we went to bed, we lashed our sledge

Chapter V.

load with extra care ; and in this we were right, as the next day we missed a saw, a knife and an axe. After a whole lot of bickering and unpleasantnesses we at last succeeded in getting these things back again. But there could be no question of leaving any depôt in the neighbourhood of these people. The first thing they would do when we were out of sight would obviously be to plunder the whole depôt. I, therefore, found it advisable to return to our friends the Nechilli and place the depôt under their charge. One day more or less to the North would not be of great consequence. On the evening previous, when the snowstorm at last ceased, we had seen land on both sides. To the west, lay Cape Hardy on Matty Island, and to the north-east, probably Cape Christian Frederik on Boothia Felix.

As usual, we travelled much faster home than out, and by 4 o'clock we were back with our good friends. Poieta got a knife for his trouble and was delighted with it. His wife received some sewing needles, and we became excellent friends. The next day we rested in the camp, as they told me they would be going south on the day following, and I wanted to see the process of removal. I did not regret it, as I never had another opportunity of being present at such a march of nomads. Moreover, I was much interested in going round the huts in the course of the day and chatting with them.

At half-past seven in the morning all was ready for starting. In all there were nine sledges, to which both men and dogs were harnessed. Many of the women were employed as draught animals, and smart they were, too, and a pleasure to look at. Not the least pathetic part of it was the good humour with which they tugged away; their faces changed alternately from red to white, and *vice versâ*, from the sharp cold and their own efforts, and I thought many of them very pretty indeed—this after barely four days' acquaintance with them. They stepped out like men, and in their gait reminded one of

ESKIMO REMOVING.   AN ESKIMO WOMAN RUNS IN FRONT TO ENCOURAGE THE
DOGS AND SHOW THEM THE ROAD.

young tars, with their swinging arms and well-bent knees. Unlike the men, all Eskimo women are bandy-legged, from always sitting with their legs bent under them. They made frequent halts to take breath, and well they might, as their loads were heavy. Between their sledges and ours there was a difference of a thousand years' evolution. Ours of the twentieth century were quite insignificant in size compared with theirs, which, like the Eskimo themselves and all that they have, belong to the stone age; and yet we carried house and provisions for three months, while the Eskimo had with them barely enough food for the day. All the sledges drove in a line, so that each smoothed the way for the next.

When a halt was made, the young folks played football. I was not able to detect any proper rules in the game, but, anyhow, it was regular football as at home, with a ball of laced-up skin, which, with the help of arms and legs, drifted about among the players of both sexes. Indeed, the women were perhaps the best players. Then the order was given for them to fall in, the ball vanished, and all were in the traces in an instant and continued their journey.

As early as noon a halt was called for the day, at a point to which, on the previous day, two sledges had been sent with their supply of meat. The Eskimo are seldom in a hurry; time is of no consequence to them, as a rule, and what they do not manage to-day they do to-morrow. Also, when pitching a camp, they take plenty of time in making preparations. The heads of families sound the snow all about with their snow sticks, and make a long and thorough examination before they decide on a site for their igloos. With good help, our hut was finished at the same time as the others.

I prevailed upon the Eskimo to remain with us and go with us to the ship on the following day. But next day they wanted to stay another day and try seal catching, and I remained to accompany them in their sport. We

started with a company of twenty men. It was bitterly
cold, with a heavy snowstorm from the north-west. I
rigged myself out in my new clothes and drew the hood
as tightly as possible over my face. The snowstorm,
which blocked our view at a few paces distance, did not
trouble the Eskimo. They knew their way, and, as the
sky was clear, they knew their bearings, which lay towards
the south-east, away from the wind. Gradually the
hunting party spread out, so that I was soon alone with
young Anguju. But while many of the others had
already found their seal-holes and begun work, Anguju's
mind was evidently intent on anything but hunting in the
intense cold. We went some distance inland to inspect
the nature of the country more closely. There was not
much to see in the snowstorm, only a little hill-crest with
projecting rocks, and I did not think it worth while to go
further. We turned round, and had a rough time getting
back: even Anguju sometimes had to go backwards
against the wind. As my nose was continually freezing,
and Anguju found it tedious to be always putting his
bare hand on it, he took off his knee-warmer, made of
reindeer skin, which the Eskimo bind round the knee to
prevent the wind from working up the trousers, and
fastened it over my nose. In this way I managed to get
back to the camp with nothing more than a slight frost
bite on my cheek. The others returned with two seals.

On the next day, March 25th, we continued our
journey together. We laid down our depôt a little way
inland, erected a high snow pillar over it, and told the

Chapter V.

Eskimo to look after it. Then Hansen and I bade them good-bye, as, with our lighter sledges, we ought to get along much faster than they. I drew the stretch of coast from King William Land in the snow, and indicated the position of Gjöahavn. They knew it well, and, like the Ogluli, called it "Ogchoktu," a name in common use amongst ourselves. Atikleura's youngest brother, Teriganyak, accompanied us, and was of great assistance, especially in the afternoon, when we had to build an igloo, as the wind began blowing again later on.

We were on board again at 8 o'clock on the morning of the 26th. Circumstances had once more prevented me from getting as far as I wished, but still it was satisfactory to have laid down a depôt so far ahead.

The day after our departure the Lieutenant and Ristvedt had set out to investigate more closely the two islands we could see ahead. That they were islands there was no doubt. In the spring the Eskimo had caught a large number of seals there, and had named these islands Achliechtu and Achlieu. (Later they were called Hovgaard's Islands.) The explorers had not yet returned. But at noon there was a lively scene, as Ristvedt and the Lieutenant arrived with all our thirty Eskimo friends, whom, to their astonishment, they had met out on the ice. Cries of "manik-tu-mi" had reassured them, and here they were all in a body, and Ogchoktu became populous. The Eskimo built themselves a row of huts in the Lindström Valley, one of the small valleys leading up from the harbour.

So much hospitality had the Eskimo shown Hansen and myself that we, of course, had to make some return ; but, at the risk of appearing stingy, I stipulated from the outset that only those who were constantly employed on board should receive food. We could not, like the Eskimo, renew our supplies by a trip out on the ice and it was, therefore, necessary to draw the line in time, nay at once. I also gave strict orders that none of the property of the Expedition should be given away or bartered. This was done to keep up the value of our means of barter, and we succeeded in so doing up to the end. The Eskimo with their sharp business instinct had soon discovered that it was more remunerative to bring their goods as gifts. I was, therefore, obliged to decline all gifts, and introduce regular trading instead. However, to show the Eskimo that good behaviour and friendliness paid, I presented Atikleura with an old Remington rifle and some few cartridges. His joy and pride were indescribable.

At this time we got some more accurate information about our good Teraiu. His whole story of the winter was a lie from beginning to end. He had purposely let the others start without him. Nor was he without food. He had six heavy reindeer carcasses hidden away among other things, in the neighbourhood of his hut—the rascal. However, we had derived much advantage and amusement in many ways, both from him and his wife Kayaggolo, so we were not too hard on him.

Meanwhile the moment came for our final start with

Chapter V.

the sledge expedition, which we had fixed for after Easter. My plan was to push on to Leopold harbour, but I have been reluctantly obliged to abandon it. Experience had shown us that our dogs were too young to manage the long day journeys we should have had to take. My first decision was to take Hansen with me this time ; but as Lund would have too much to do on board alone, I had to give up Hansen and take Ristvedt instead. Such changes in plans easily occasion dissatisfaction, and are therefore unpleasant. But there was no help for it in this instance. On April 6th we were ready. We went with Teraiu and Kachkochnelli, who were going with their families to Abva (Mount Matheson) to catch seals. The day was fine and, with its 22° below zero, could still be called a spring day. This was the first day in the year that we felt the sun's warmth, an inexpressibly pleasant sensation. Soon we had to take off our outer clothing, lay it in the sledge, and continue our journey in undergarments only. Kayaggolo—" the old Eagle "—had gone on a couple of hours in advance to get a start. But the poor thing mistook the way out on the ice, and had to go a long way round to get into our line of march again. But it was tedious travelling for us too, for fresh snow had fallen and made the road heavy ; the sledges were heavily laden, and the dogs out of practice. Consequently, Kayaggolo reached the land on the other side of the bay behind Neumayer's Peninsula at the same time as we did. But she was very tired, the poor creature.

While preparations were being made for lunch I struck up one of my best tunes, and the " Eagle " revived, and joined with all her might in a duo, which in other latitudes would scarcely have been so well appreciated. The menu consisted of the daintiest frozen dishes—reindeer meat, fish, and reindeer tripe, garnished with blubber chopped up in squares. Ristvedt is no despiser of good fare, and he swallowed piece after piece of meat and blubber. It is an invaluable quality in a man on such an expedition to be able to eat anything. Lieutenant Hansen, too, had a most accommodating appetite; at a pinch he would have made a meal of nails and pebbles. But at the existing temperature, however much we might call it spring-like, we could not stop long for refreshments. The whips were soon cracking again, and the dogs and men were off once more. The old " Eagle " was invited by the ever-polite Ristvedt to seat herself on his sleigh, and so she was very well off.

At half-past six in the evening we bade farewell to the Eskimo near our depôt. They were going on to the nearest camp, while we intended to spend the first night here. The separation would not be for long, as they were to return to Ogchoktu for seal catching. With some anxiety we examined the depôt, and found it untouched and in order, to the great credit of our friends the Nechilli, to whom the wood and iron materials stowed away would have been immensely valuable. And it would not have been difficult for them to steal the whole of it, hide it till we were clear away, and then enjoy the benefit of

Chapter V.

it. The tent was now sufficient as a night shelter, as the temperature seldom went lower than 22° below zero. Small as it was, it rendered valuable service, and we saw it fixed up with delight as a sign that the day's toil was ended and the night was before us.

At 8 o'clock next morning we went on ahead. We had loaded 600 lbs. on each sledge; with strong, well-grown animals, five dogs for each would have been ample; as it was, the oads were too heavy and our progress was slow accordingly. We also felt the want of the Eskimo, who on the day before had driven in front and smoothed the way for us. From Abva, where the depôt lay, we shaped our course towards the ice and then southwards to Matty Island, where I wished to place my first magnetic station. In the evening, when we were about to select a site for our tent, to our joy we discovered an Eskimo camp, a good distance away, with at any rate one living inhabitant, a solitary black dot moving among the huts. We made a rush for the igloos, but, alas, found an abandoned camp. The inhabitant we had seen flew screeching up behind one of the huts. It was a raven.

In the evening of April 9th we reached Cape Hardy, on Matty Island, after a hard struggle over the pack ice, which blocked the coast; we had to leave one sledge behind and harness all the dogs to the other. Next morning, at half-past six, we took the dogs with us in leashes and fetched the sledge we had left behind. The whole distance was nine miles. Later in the day I took our latitude and longitude, and by the time Ristvedt, who

was an excellent cook for such a journey as this, had
served dinner, the observatory was built. It was simply
a round wall, just high enough to shelter me and the
instrument. From the observatory to the tent, which
was about 200 yards away, I stretched a cord, making
the other end fast to Ristvedt's arm. By this means he
could, while lying in the sleeping-sack, read off his watch
every time I pulled the cord ; this was very convenient
for me, as I was thus spared the trouble of recording the
time. Ristvedt was hugely delighted each time he got
a tug.

I took observations from 5 o'clock to 6 o'clock in the
morning and at night. The temperature went steadily
up. The first night's minimum was − 16·6° Fahr. One
afternoon when I was going to read the time, I unfortunately slipped and broke the watch-glass. Then we
emptied the pepper caster and used it as a watch-case.
On the whole the work went well, and by April 14th we
had finished. On the 15th we went on further. There
was a thick fog, and, just as we halted, two Eskimo
emerged from the gloom. They belonged to the same
gang that Hansen and I had met in March. But, under
present circumstances, we had to be on friendly terms
with the two gentlemen,—Kaumallo and Kalakchie.
They found their way easily through the dense fog and
we were soon at their camp. Their hut was the only
one left, and these two, with an old woman and two
children, were the only occupants left in the bandits'
camp. They had evidently repented of their previous

Chapter V.

behaviour and were now very courteous. We put up our tent near their hut. A storm from the north with thick snow prevented us from journeying further the following day. Meanwhile Ristvedt harnessed all our dogs to one sledge and went south to the vessel with Kalakchie to get the watch mended or changed so as not to waste time. The distance there and back was 108 miles, and on the 20th at half-past seven in the evening he returned having satisfactorily accomplished his task. To remain four or five days in idleness, with an old Eskimo woman, a man, and two children, was not diverting, and it was therefore refreshing to get away again the next morning.

The ice off Matty Island was awkward, and to get ahead we had to leave one of the sledges behind. We, therefore, drove ashore on to Boothia Felix and laid down a depot a little northward of Cape Christian Frederick and left the other sledge there. We then went back to the first sledge and passed the night there. We had done twenty-four miles that day. Next day we proceeded through the hummock ice with all the ten dogs put to the sledge, carrying a load of 660 lb. On the north side of Matty Island we came upon an Eskimo camp of three huts and pitched our tent with them. Here we also availed ourselves of the opportunity to give our dogs a good feed. Ristvedt had a good meal, his favourite dish, seal meat fried in oil. I am not dainty, but I would rather be without it. It was among these Eskimo that we saw for the first time little Magito, who afterwards became the belle of Ogchoktu. She was twenty years old, married, and very handsome ; I was not the only one who thought so.

The journey up along the coast of Boothia did not present any interesting feature. We were very near

OUR TENT CLOSE TO THE MAGNETIC NORTH POLE.

the Magnetic Poles, both the old and the new, and probably passed over them both. We took up our northernmost station a little to the south of Tasmania Islands and turned back on May 7th. My intention was to go back, fetch our depôt, and with it make for Victoria

Chapter V.

Harbour, where the two Rosses had wintered with the "Victory" in the thirties of last century. A series of magnetic observations here would be very interesting, possibly even more interesting than at the Pole itself. However, I was not to succeed in carrying out this plan. When on the trip south we passed the old winter-quarters, my left foot, which had been troubling me for some time, probably as the result of tight lacing on the ankle, became so utterly useless that I had to lie up. I was in this condition from the 12th till the evening of the 18th when we were able to go on. Fiks and Syl, the two inseparables, had disappeared during a bear hunt we had had a little further north on the coast, and were never seen again. Ptarmigan showed themselves from time to time, and Ristvedt, with his fowling piece, procured us more than one splendid dinner.

The land along the whole of that coast of Boothia is as flat as a pancake. Far inland rise a few peaks, visible some distance out on the ice.

On May 21st we reached our depôt, but found it entirely plundered by our friends Kaumallo and Kalakchie. Eleven pounds of pemmican lay scattered around, that was all. We had no choice but to set our course back to the "Gjöa" as fast as we could, as these ten cakes of pemmican, with a couple of packets of chocolate and a little bread, represented our entire stock of provisions. A strong northerly breeze helped us along, and we travelled south at a steady and brisk pace. The weather was thick, and after a couple of days we found we had strayed inland. But then it cleared up, and on the evening of May 27th we were on board.

So our trip was not a brilliant success ; but, considering the many untoward circumstances which had occurred, we had to be satisfied with the results.

CHAPTER VI.

Summer.

On board they had had a raw and cold May. In spite of various harbingers of spring, no one would ever have imagined, even at the end of the month, looking at the " Gjöa," buried as she was in snow and with steep snow-drifts reaching half way up her masts, that spring was approaching. During the whole time, the health of those on board had been very good, and they had received visits from Eskimo from all the four corners of the earth. Lieutenant Hansen and his assistant Helmer Hansen had carried out their cairn building very thoroughly ; on every summit and knoll the cairns could be seen and a good deal of the Neumayer Peninsula with the permanent base had been charted. Lund had continued his great struggle with the snow in order to keep the vessel free. In the month of June the ice in the harbour was 12 feet 6 inches thick, so he must have had plenty of work to do to keep the water hole open. Wiik had taken excellent magnetic observations ; Lindström was fat and contented and cooked better than ever. He also showed considerable aptitude as a hunter, and many a ptarmigan fell a victim to his gun. This latter achievement was far from pleasing to Lund and Hansen, as they were the professional huntsmen of the party, so one morning they determined to play a trick on the sporting cook. They quietly fixed up a frozen ptarmigan, killed some two months previously, on the top of a snow drift out on the ice about twenty-five yards away from the ship. Lund went down to the fore-cabin where Lindström was still at his breakfast and called out, " Lindström ! Lindström ! there's a ptarmigan yonder on the ice." Like a flash of lightning Lindström appeared on deck with his gun. " Where is it ? " " There, on the bow." With stealthy steps and crouching in approved fashion Lindström went cautiously forward. Hansen was looking over the rail, and Lund close behind. The ptarmigan seemed to be asleep in the snow-drift. Now, one would imagine that a ptarmigan asleep, just under the boat, would have awakened some suspicion, but Lindström regarded it as quite natural that a bird should be taking a morning snooze, so he took up a good position, aimed and fired. The ptarmigan rolled over and over on the ice. " Ha! Ha! I hit him that time," cried Lindström proudly, as he sprang over the side and ran to fetch his prey. For a time he stood with the bird in his hand feeling it all over. " Why, it is quite cold," he cried out in astonishment ; but from the peals of laughter which immediately broke forth, Lindström saw he had been tricked.

On June 3rd the Eskimo began to return for the seal fishing on the ice. They settled on Von Betzold Point,

Chapter VI.

from which they had the most splendid view over the whole fishery. They brought us a quantity of blubber from seals they had caught in the course of the winter, and we took all we could get, giving wood and iron in exchange. It is impossible to tell what one may need in the future. Instead of stowing away every specimen as they had been in the habit of doing in winter, they sold them to us and our collection was considerably increased.

On June 5th we started on a combined magnetic and surveying expedition. The Lieutenant, with the assistance of Hansen, had to survey from the cairns on Achliechtu and Kaa-aak-ka, and I, together with Ristvedt, had to establish magnetic stations on the same places. We took provisions with us for fourteen days, and also the Eskimo Ugpi and Talurnakto, who, later on, were constantly employed by us. Ugpi or Uglen (the " Owl ") as we always called him, attracted immediate attention by his appearance. With his long black hair hanging over his shoulders, his dark eyes and frank honest expression, he would have been good-looking if his broad face and large mouth had not spoilt his beauty from a European stand-point. There was something serious, almost dreamy, about him, and he was quite free from that very annoying and wearisome custom common to other Eskimo of always making fun of the failings of others. Honesty and truthfulness were unmistakably impressed on his features and I would never have hesitated for a moment to entrust him with anything. During his association with us he became an exceptionally clever hunter both

THE "OWL" AS ARCHER.

for birds and reindeer. He was about thirty years old, and was married to Kabloka, a very small girl of seventeen. They had no children. Kabloka was too

KABLOKA.

much of a Mongolian to be handsome but she captivated everyone with her child-like innocent ways. The "Owl's" mother Anana, already referred to, lived with him; but Umiktuallu, his elder brother, ruled the whole family. He was a gloomy, unprepossessing fellow, but a clever seal catcher. His wife Onallu was a fine woman but a dreadful scold.

UMIKTUALLU.

Talurnakto, our other companion, was the very opposite of the "Owl." Amongst his own people he was regarded as somewhat of a fool, but in reality was the

cleverest of them all. He was always laughing and
making a noise, had no family, except one brother, and
was daring enough for anything. Fat and small, he
was known by the name of Takichya : he who walks

ONALLU AND SON.

heavily. From the first our attention was attracted to
him by the stubborn persistency with which, after the
lapse of a few days, he regularly returned to the ship
after we had told him in a friendly manner that we had
no further need of his services. One fine day Talurnakto

Chapter VI.

came again on board, and, as we had found out that he was a good worker, we determined to keep him. Although he had not, perhaps, the same honest expression as the "Owl," yet there was never any cause for complaint; indeed, any one of us might have been satisfied to have as good a conscience as Talurnakto. He soon became so accustomed to the food on board that he got quite fond of it. But his "manners" were dreadful. He was an awful fright to look at, and crammed his food into his mouth with whatever he could lay his hands on. On the other hand, the "Owl" was an example in this respect. In evening dress he would have conducted himself irreproachably in the best society, and would have handled a knife and fork like a perfect gentleman.

We started with these two companions at 10 o'clock in the evening; the snow was as hard as stone, and strong enough to bear anywhere, but, as my left foot was still very painful, and could not stand the unevenness of the ground, I went on ski. It was a little too warm to drive in daytime. The dazzling light was so trying to the eyes that even with the darkest glasses one risked becoming snow-blind; but the pleasant weather cheered us all. The dogs were fresh and strong, and we started off at a good speed, singing and laughing. By 12.30 A.M. we had covered about twelve miles to Achliechtu. In the month of March the Lieutenant and Ristvedt had already ascertained that this was an island. M'Clintock, during his expedition of

TALURNAKTO.

1859, passed very near to this place—a few miles to the east of it, but the continual bad weather which dogged him at the time prevented him from observing it. As soon as we got on the island we found some bare spots, and took possession of them, for erecting the tents. One must have lived in regions where everything is covered with snow for some ten months to understand the enthusiasm occasioned by the sight of these bare spots of earth ; we stamped around on them with a wonderful confidential feeling of security of being once more on our Mother Earth ; our pleasure at the sight of earth and moss was as if we had just met an old and faithful friend who had been long absent. We could not exactly say it was very spring-like here as yet. But the awakening life, as compared with the deadness of the long winter, had quite a cheering effect. The few bare spots were like small worlds in themselves, swarming with humming insects ; a little flower peeped forth, and was hailed with joy ; ducks, geese, and swans continuously passed high above our heads towards the north in large flocks. We shot a great number of ptarmigan, and lived happily on fresh food ; altogether, the days we passed on Hovgaard's Islands must be included amongst our happiest. An igloo, which our two Eskimo had constructed for myself and my instruments, quite surprised me one day by tumbling down, and burying both the instruments and myself, fortunately without any injury to either. Every now and again we were visited by Eskimo, whose camp we could see a little distance away on the ice. The surveyors went over to the other island, and found it very long and low.

By June 10th all the work was finished. Hovgaard's Islands were charted, and the magnetic conditions investigated. We therefore packed up our things and went on to the next station, Kaa-aak-ka. This place we were already well acquainted with, and I shall always remember my first visit there, to the igloo with the Eskimo. Now in June the place had still quite a wintry aspect ; only on the sharpest and highest " ridges " had the snow melted, every other part being covered with it. The Lake of Kaa-aak-ka was also ice-bound ; in the holes for fishing, the ice attained a thickness of 10 feet 6 inches. We only found just enough clear ground for erecting tents. Yet even now one could see that this place must be a real Arctic paradise in the summer, when the great expanse of water lies open and glistening, when plants and flowers cover the hills, and the whole place is enlivened by the presence of a multitude of happy birds. It must be also an ideal region for reindeer. The day after our arrival our Eskimo shot two reindeer, the first we had seen. There were also a great number of ptarmigan, of which we shot many. The Eskimo hunted early and late. The survey department was in constant activity ; we saw them wandering from one cairn to another, and now and again heard the report of a gun in the hills, and knew that they had encountered game. I myself was very much occupied with my observations, and enjoyed the work, but poor

Ristvedt suffered all the torments of Tantalus; this passionate hunter had to keep to the tent the whole time; I required his assistance with the observations, and, besides, he had to do duty as cook. It was very hard for him to hear the report of the guns in the distance, and even harder, perhaps, when the others returned home with splendid bags, to listen to their enthusiastic stories of the chase.

The surveyors preferred working during the night, and we called them " night-birds," and they in turn dubbed us the seven sleepers of " Ephesus." Each party kept house for themselves, and there was always some dispute between Ristvedt and the "night-birds'" cook. Ristvedt was really an expert chef in the culinary art, and was a great gourmet as well. Now the survey party on their excursions obtained quantities of fine, fat ptarmigan, but we had to content ourselves with the few poor thin ones the Eskimo brought us; and it happened more than once that Ristvedt went to the other larder and changed birds, to the great annoyance of the other cook.

The Eskimo were very useful to us; they brought us reindeer every day, and when we started on our homeward journey on June 14th we had a large store of reindeer meat with us, to the great delight of those on board whose stock of fresh meat was exhausted. We arrived at half-past three in the morning. Ogchoktu lay in the deepest sleep. We immediately noticed a great change since our departure; the " Gjöa " had

Chapter VI.

thrown off her winter cloak ; the covering over the ship had been removed ; below deck, too, she was free from the yoke of winter. Lindström had indeed worked splendidly. He had got rid of the large masses of ice, previously to be seen like glaciers in every nook and corner on board. All the walls were bright and clean, having been washed and cleansed from the winter's soot and dirt ; the many double skylights with their deerskin linings were gone, and light and air had free access ; we drew a deep breath of freedom as we came on board. It was spring.

The Eskimo came to the spot in large numbers, and settled down around us. The cleverest and best of them had raised their igloos and made them cosy. The laziest, on the contrary, used the old huts, which, in reality, were not huts as we generally understand them, but holes under the snow, into which they crept through an opening in the top. Others built a round wall of snow and stretched a roof of skin over it. They thus risked the unpleasant surprise I had twice experienced of suddenly getting the roof as a blanket.

I previously mentioned that we had found Eskimo morality very excellent ; even stern critics could not have found fault with their conjugal morals. Now, on my return after ten days' absence, I found the most astonishing change in their respect. All shame was cast aside, and men offered their wives, old cronies, their daughters or step-daughters for sale like any other merchandise. I was never able to find out the reason of this alteration, but from this time onward we never said much about morality among the Eskimo.

The anniversary of our departure from home was kept as a holiday. The next day Ristvedt and I started off again to establish a circle of magnetic stations around the head station, for investigating the magnetic conditions near the base stations. We began our journey at 10 o'clock in the evening with our two Eskimo, and took the direction of Tyataa-arlu (Point Luigi d'Abruzzi). As the snow was not fit for building purposes, I took my observation tent with me. The Duke of the Abruzzi's Point is a very low, flat, sandy point, stretching out in a southerly direction from Abva (Mount Matheson). It was overgrown with moss, and full of small lakes, a veritable paradise for birds. As we were setting up our tents in the early hours of the morning, two ptarmigan, regardless of the mighty Nimrod of the expedition, came right in upon us and began to fight, their feathers flying in all directions. They were very soon compelled to transfer their battle-ground to our soup-pot.

As I became accustomed to these field observations, I learned to dispense with assistance ; so Ristvedt, much to his joy, was able to devote himself almost wholly to hunting with the Eskimo. These three hunters secured an abundance of game and we used but little of our own food. The best dish of all was reindeer's tongue, which almost "melted in the mouth." Eider ducks just at this time, when they are migrating, are plump and of a very fine flavour. The fat meat round the seal flippers was also excellent, especially for bouillon. When the meat is prepared in this way it does not taste of fish oil, but rather reminds one of mutton fat.

SUMMER SCENE ON DECK. GJÖAHAVN, 1904.

After having terminated the observations, we left Tyataa-arlu and set our course inwards to the head of Schwatka Bay. This bay runs ten miles inland. It gives one the impression of being very dirty; the upper end or head of the bay is quite blocked with islets and rocks ; if a navigable course could be found in the bay, this end would offer a splendid wintering spot, as the region around is very rich in game. From Schwatka Bay we continued N.N.W. towards the high and easily recognisable summit we had made our goal; this we named " Nordligste Nordhöi " (Farthest North Hill). The snow was now loose and the Eskimo were up to their knees in it ; the stout little Talurnakto, especially, found it hard work. Ristvedt and I went over it easily on our ski. The whole of the way up was covered with reindeer and fox tracks. The ridge had begun to assume its summer garb ; large stretches were quite cleared of snow, and many small lakes were unfrozen. In the bright sunshine, day and night, might be seen swarms of eider ducks, swans, and loons, and the little birds chirped so unceasingly that you wondered whether these beings ever slept during the twenty-four hours. There were also numbers of geese. But the birds were very wide awake and shy, so that those we killed only sufficed to supply our daily needs. One day the " Owl " brought home four lemmings ; we gave these to the Eskimo. They ate them as though they were the daintiest fare, but told us that later in the year, when the lemmings were too fat, they did not care for them. The ground about us swarmed with these animals, and it was one of our greatest amusements to sit outside our tent of an evening and watch these funny little creatures ; their holes were all round us. They came out to bask in the sun and heat. Then they disappeared and we heard cries and screams of domestic strife below ground. Again they would come out for a little exercise, rolling themselves along like balls; they are as broad as they are long, and their movements and habits have something indescribably funny and amusing about them. If they met a dog or a man they would sit up defiantly on their hind legs and snarl and hiss, as though eager for the fray. Poor little creatures, they had many enemies, especially the owls. On one occasion we found six of them dead in an owl's nest; that was one for each of our dogs. There were many of these loathsome birds of prey. One day Ristvedt shot at one but only injured it so that it could not fly. It sat down and waited for us; but its daring, wicked look and horrid appearance quite frightened both ourselves and the Eskimo, and we finished it off with another shot.

The flowers and herbs were now sprouting, and millions of insects hummed and buzzed and fussed busily around us after their long torpor. All this naturally rendered these sunny days doubly attractive to us, and we were so happy there that it was with regret that we left the spot. Our next station was "Nordhögda"; it was very beautiful, with its steep sides looking south, west, and north; towards the east it formed a small pebbly ridge, surrounded by lakes. But our most beautiful camping place was on Wiik's Hill, from which we had the best and most extensive views. At the foot of the hill lay a lake full of fish; this we named "Great Ristvedt," and this, with another we called "Little Ristvedt," communicated with Ristvedt River. Our stay upon Wiik's Hill was the crowning feature of this excursion. Beautiful sunny weather every day brought the temperature in the observation tent up to 77° Fahr., so I had to dispense with all my heavy clothing and only retain what was

SUMMER ON KING WILLIAM LAND.

absolutely necessary. The Eskimo were always on the move and brought home large supplies for the larder. Here at Wiik's Hill we also found a great quantity of eider, loon, and geese eggs. All these eggs were sent on board for Lindström's collection. It was rather hard to part with them in this way, but as we found comparatively few eggs, the collection had to take precedence of our desire for delicacies. During the great heat it was very pleasant, after finishing our work, to take a bath in the lake or river. Ristvedt would then lay the cloth for supper, and we would enjoy our well-earned rest. One evening Ristvedt came over from the Eskimo tent and told us he had refused an invitation to supper there. As he stated that the reason of his refusal was that the food they offered him had even exceeded the limits of what he regarded as eatable, I was very anxious to know how it was prepared. It was a new sort of blood-pudding. Ristvedt was very fond of blood-pudding, and one of his specialities as a cook was something he called blood-dumpling ; this was really very good. But the Eskimo pudding, well . . . . no thanks, he could not manage it ! He had followed the preparation from the moment the deer was shot. It was at once skinned and the blood carefully collected. A portion of the tepid blood was drunk up by itself, then the stomach was taken out of the animal. The Eskimo partook of a portion of the contents by scooping it up with their hands. When the stomach was half empty, they put the blood into it and stirred it round with a thigh bone. The dish thus prepared was blood-pudding *à la* Eskimo, which even Ristvedt had refused to partake of. After they had eaten a portion of the fresh " pudding," they tied up the stomach and put it in a spot exposed to the sun, covering it over with a flat stone ; there it lay to "season" until late in winter ; it was then ready and " ripe," and was used for banquets.

Just below Wiik's Hill we made the interesting discovery of a skeleton of a whale, whose vertebræ projected from the hill.

On the very day we left here it began to rain; this, in fact, was the end of summer. The weather continued mostly cold and windy for the rest of the time. At the next station—Adolf Schmidt's Hill, or Saint John's Hill, as it is now called—my good assistant had to leave me to go on board and put the engine in order. One never knows what may be in store, so we had to be prepared for all contingencies. We had the sails bent the whole time we were in the ice, and this, indeed, proved the best method of preserving them.

I was now barely a mile from the bay, and in my unoccupied hours I generally took a turn on board. Here everything was in order. The Lieutenant had completed his triangulation of the bases. Lund and Hansen had to look after the boat and keep it in repair, and they did this work very thoroughly. Wiik conscientiously carried out the magnetic observations. After the charting work was finished, the Lieutenant offered to increase our zoological collection. We were all interested in this, and got a great deal of material together; the Eskimo also helped us considerably in completing it. Lindström gave prizes, principally consisting of old underwear, for which the Eskimo competed eagerly. Later on we met Eskimo on King William Land strutting about in Lindström's worn-out pantaloons, etc. Altogether, he worked indefatigably, and he endeavoured to obtain a specimen of every living creature in the region. Even the special

Chapter VI.

kind of *pediculus capitis*, which the Eskimo rejoiced in, had to be obtained, and Lindström offered a prize for specimens. At first very few specimens were obtainable, but when the Eskimo understood that it was really a business matter, they came daily in crowds, bringing specimens to Lindström. What had previously been Lindström's joy was now his despair, and it required all his energy to put a stop to this business and keep the *pediculi* at a distance, but even then he had enough of all varieties to furnish a good supply to every zoological collection in Europe.

The ice on the more extensive lakes now began to break up, and the Eskimo caught plenty of trout and brought them to us. The ice in the bay was crowded with fishermen, and we made arrangements with one to bring us small fresh cod whenever we required them. We generally had them fried for supper. Reindeer beef was almost invariably served for breakfast. On the whole, the life around us was very pleasant to contemplate. Little dots of five to six years sat with their fishing lines, often through the whole night, and made very good catches, sometimes more than their own weight. The youths and the men undertook the fishing on the lakes. They fished in the small leads close in to the land, where the water was so clear that one could follow every movement of the fishes.

Under such circumstances, all of us, as well as the Eskimo and our dogs, lived contentedly, and had no concern as to food. Yet it was very clear to us that the

Eskimo are taught not to allow anything eatable to be wasted, although their idea as to what was eatable differed considerably from ours. When we were stationed on Adolf Schmidt's Hill, old Eldro, or Praederik, as he liked to be called (he really meant Frederic), was sitting one morning outside my tent, and as I came out I could see he had something on his mind. He prattled away for some time, but I was unable to understand a word of it. By dint of time and mutual intercourse a sort of Norwegian-Eskimo dialect was developed, by means of which we could easily understand and be understood by the younger and more intelligent Eskimo ; but Eldro, like a great number of the elder ones, adhered purely to Eskimo. He kept on talking so long, however, that I at last began to perceive that he was asking me to give him something of mine that was on the hill. I thought he might be referring to a piece of wood or metal that had been thrown away, and granted his request, whereupon, beaming with joy and expressing his thanks, he went off ; shortly after I went down and saw old Eldro sitting on the slope, busily engaged on something. Being curious, I approached him, and found him carefully collecting the contents of an old reindeer paunch which had been lying there since the previous autumn. My dogs had been down and sniffed at the filthy stuff, but, rearing on their hind legs and turning up their noses at it, they had left it severely alone. But Eldro smiled happily at his unexpected windfall, and told me that his wife would be delighted at this addition to her larder,

Chapter VI.

I murmured something that he shouldn't be too sure, ladies were so peculiar. The next day, however, the old gentleman presented me with a couple of dozen of fine trout, and thanked me gracefully on behalf of his wife for the food I had given them.

After having made two stations, the one on Sunday Hill and the other on Swan Hill, I terminated my summer expedition for the present, as the ground was very bare of snow, and travelling in the sleighs was therefore far from easy. On the evening of July 18th I was again on board.

The previous evening a very tragic event had occurred at the station. Umiktuallu, the "Owl's" elder brother, to whom I have previously referred, lived with his wife, three children, and a foster-son, in a tent pitched a few paces below the "Magnet." He had in his possession an old muzzle-loading rifle he had obtained by barter from another Eskimo. He had procured balls, powder, and caps from us. He was accustomed to leave the weapon loaded, which indeed in itself was not very dangerous, but in spite of our repeated advice he had not removed the caps. That evening, when he and his wife were visiting another family, his foster-son and his own eldest son got hold of the rifle. Then followed what so often happens when boys play with weapons without having been shown how to use them properly ; they were ignorant of their danger, the gun went off, and Umiktuallu's son, who was only seven years old, fell down dead. The father heard the shot and rushed

UÄIKTUALLU KILLS HIS FOSTER-SON.

to the spot. At the sight of his own dead son, and the foster-son sitting with the smoking weapon, he was seized with frenzy. He carried the horror-stricken boy out of the tent, stabbed him three times through the heart with his knife, and then kicked him away. Wiik was a witness of this terrible scene from the " Magnet." The seven-year old lad was an exceptionally bright and clever little fellow ; he was really quite a hunter, and with his bow and arrow brought quantities of game to the house. Umiktuallu was exceedingly fond and proud of him. Both boys were buried that night, we did not know where. With time and reflection Umiktualla calmed down, and was seized with remorse. When I entered the camp the next evening he and his family had gone over to the mainland.

As usual I found everything in the most perfect order on board, and I put up my observation tent in Ogchoktu so as to go through my magnets at the station. Now the ice in Simpson's Straits gradually assumed the bright blueish-green colour seen when the snow melts on the surface ; probably it will not be long before it breaks up. The Eskimo told me that it breaks up every year, but they also informed me that the summer of 1903, in which we arrived, had been a very exceptional summer for ice, and that I must not expect a repetition of it. However, our prospects were of the best. The spring and the early summer were marvellous with their long warm evenings, in which we could enjoy Nature's re-awakening to life. There was one pest, however, which was bad

Chapter VI.

enough—the gnats. They were so bad that it was almost impossible to remain out in the open air, and we fought and battled with the swarms as though they might have been hoards of raiding bandits. They even followed us on board, and in order to get any rest at night we had to cover over all the skylights with mosquito netting. The stings of the knats had but little effect on me, but the others were more susceptible to the poison, especially Lund, who used to swell up terribly, and his deformed features gave us a fresh surprise every morning.

We had set our fishing nets in one of the many lakes, and kept ourselves constantly supplied with trout; we took turns in attending to and dragging the net, and there was constant rivalry as to who should bring the most on board. One day Lund and Ristvedt were out on the lake and caught two trout, a large one and a very small one. The came slowly home with this rich capture, knowing that they would be the victims of all the sarcasm and chaff the " Gjöa " could muster. But suddenly Lund interrupted the brooding silence and mentioned the name " Lindström ! " " Why, of course," said Ristvedt, " that's the man."

Not a word more was spoken but that was enough, and for an hour they went on in silence both pondering over Lund's idea of turning the laughter against our well-fed cook. While they were away with the boat the plans were laid. It was three o'clock in the afternoon and the cook was enjoying his happy siesta when the two fishermen burst in upon him. " Lindström ! Lindström ! Hi

ADOLF LINDSTRÖM WITH SAMPLES OF FISH ON KING
WILLIAM LAND.   AUTUMN, 1904.

—hi, Lindström!" Lindström sat up in his hammock
and pushed his head through the curtains. "What's
the matter?" "We have come from the nets," said
Lund. (As Lindström always mistrusted Ristvedt, Lund
was put forward to do the talking and so disarm any
suspicion), "and amongst what we have caught we have
found two specimens you might like to have in your
collection." Lund and Ristvedt each stood with his fish
in paper holding them with much care as if they were
babies. They carefully unpacked them and proudly
showed them to Lindström. "See here," continued
Lund, "this is the mother and that is her young one."
The rascals knew Lindström, and also knew that as a
zoologist he attached great importance to having specimens both of parent and offspring. "Ha!" said he
eagerly, and his round features brightened at the thought
of what Professor Collett would say when he came with
the mother and young of a fish family. He sat right up
in the hammock and seized the little fish. His quick
glance of suspicion was so frankly met by the two friendly
fishermen that he was reassured, and with the expert air
he always assumed whenever collections were discussed
he began to examine the specimens. He used some
zoological terms, speaking rather to himself than to the
others, and when his preliminary examination was over,
he expressed his thanks for their gift. "Yes, it is really
most interesting, because this little one, which must be
a year old, must have followed its mother about during
its first year of existence."

Chapter VI.

He already seemed to see his theory confirmed by the fish he held in his hand, but the roar of laughter that came from the fishermen sadly convinced him that they at any rate did not accept his hypothesis. They had achieved their purpose brilliantly and Lindström was the " butt " for the rest of the day.

CLOSED NECHILLI GRAVE.

Most of the Eskimo had now left us and returned to their summer dwellings for reindeer hunting and fishing. Only three families who intended going north to King William Land remained behind ; besides these we retained the " Owl " with his mother and wife as well as Talurnakto, as they were to accompany me when I resumed my interrupted observation journey as soon as the leads were practicable in a boat. For several days we searched for the bodies of the two Eskimo boys as we were very curious to find out how they were buried ; at last we were lucky enough to find them ; each lay in

AN OPEN NECHILLI GRAVE CONTAINING THE BODY OF THE BOY
MURDERED BY UMIKTUALLU.

his own grave surrounded by small stones on the hill side not far from the " Magnet." The son was carefully sewn up in deerskin and buried with his bow and arrows, drinking cup, gloves and so on, but the foster-son was treated very indifferently ; his head was almost uncovered and only a pair of old worn-out gloves were buried with him. The insects had already begun their work upon him. In the course of the winter the foxes would complete it. When I visited the place a year later both the graves were swarming with small worms.

Lieutenant Hansen took advantage of the most favourable period to hurry on the development of his photographic plates. Now he had a plentiful supply of fresh water, previously so scarce, and by damming and diverting, he arranged quite a dam in one of the watercourses, and in this artificial rinsing-place he splashed and washed to his heart's content.

On August 1st I started again on my observation trip round to the station. This time I went by water, accompanied by quite a flotilla. One of our small oak rowing boats was the flag-ship with myself as sculler. Anana and Kabloka sat in the stern-sheets. All our food was also on board. Talurnakto pulled one of our canvas boats, and the "Owl" was in his own kayak. The kayaks of the Nechilli Eskimo are clumsy and plain as compared with those we saw amongst the Eskimo of Greenland, but these latter are much more dependent on their kayaks for getting from place to place. Our destination was Helmer Hansen's Hill, not quite five nautical miles away, so we had not very far to go. The gnats were very bad until we got well out on the lead ; here they left us. However, as it cleared up in the course of the day after the morning's rain, and the sun came out very strong, we got most terribly tormented

on landing. The half hour's walk from the shore up to
the cairn was hell upon earth. We walked through a

Amundsen.   Talurnakto.   The "Owl."
BOAT EXPEDITION TO INVESTIGATE THE MAGNETIC CONDITIONS ON KING
WILLIAM LAND.

cloud of gnats, and as our hands were filled with parcels
we were at their mercy; if we opened our mouths to say
a word we immediately got them filled with gnats. We

Chapter VI.

were driven to desperation, and almost decided to turn back; but we struggled on, put up our tents, and found shelter from the tormenting millions.

My stay on Hansen's Hill was very pleasant, the landscape was beautiful, and there was an extensive view. The lead reached just up to this point; to the west the ice still lay close to the land. Towards the east one looked over a large bay right into Ogchoktu. To the south there was an open view over Simpson's Strait, and to the west and north lay the endless moss stretches of King William Land, with lake after lake shimmering inland. The birds were flying about there in thousands; here and there a solitary reindeer could be seen grazing. I could hardly work of an evening owing to the gnats, and I spent my mornings and spare time in pleasant intercourse with my Eskimo friends, whom I now knew so well. The " Owl " and Talurnakto were deer hunting most of the time. Kabloka often accompanied them to help carry the meat home. Old Anana, however, remained in the tent, except for little excursions to collect heather for fuel. I also got her to help with the house-keeping. Her principal work consisted in cleaning the meat, which the Eskimo handle very carelessly; they let it lie wherever it falls, and the consequence was that it got all dirty with the hair of the deer, dirt, and pebbles. Anana's cleaning principally consisted in licking the meat, and when I tried to dissuade her from so doing she evidently thought it was out of consideration for her. Now and again, when we were all together and I had an hour free, I invited my friends into the tent and treated them to hard bread and chocolate. Of all drinks, chocolate was the best they knew of. A smile spread over their entire features at the mere mention of the name. During these chats I would ask them all sorts of questions about their life and their views on things in general. I also endeavoured to become more fluent in the language. The "Owl" was a regular pedagogue. He was greatly interested whenever I wanted to know what this or that was called. Talurnakto, however, was impossible. He simply laughed and regarded it all as the greatest fun. My own camp outfit for eating purposes was not very extensive, and I told them that they would have to manage as best they could. This they did, and showed themselves to be very practical and resourceful. One evening as we sat eating together, Talurnakto was seized with a violent itching in the back. With his short arms he could not reach the spot, and resolutely seized hold of my spoon, which I had laid down for a moment, pushed it down his neck, and scratched away.

When I had almost finished my observations, the Lieutenant and Helmer Hansen came with a dory which they had fitted up for a long excursion to Cape Crozier, the westerly point of King William Land. Here they wanted to establish a depôt for the next sledge tour in the spring of 1905, to the east coast of Victoria Land. At the same time they proposed to sound the narrowest portion of Simpson's Strait between Eta Island and the coast. They had provisions for a month, and so with what they were taking for the depôt, they were well loaded. The distance between Cape Crozier and Ogchoktu was 100 nautical miles. The ice, which up to the present had kept close to the land, was now driven out by a strong north wind, thereby opening a navigable lead. They stayed the night with me, intending to accompany me the following day, as I was going on to Kaa-aak-ka, my next station.

That afternoon I was going to take my first trip in a kayak. I chose a suitable little pond for practising on. At first I managed splendidly, but my comrades, who were building a cairn close by, made remarks that were anything but flattering. I calmly turned round to tell them that I evidently had a natural talent for rowing a kayak. At that instant both the kayak and I turned over. The water was so shallow that I touched the bottom with my arms, but the kayak filled and I was wet to the skin. The others had to drag me ashore and my return to the tent to change my clothes was not a march of triumph.

Next morning we went westward in the dory; the wind being favourable, there was no need for rowing. Anana, who was not found of the sea, preferred to go by land; the "Owl" was in his kayak. Kaa-aak-ka was now in full summer garb; a variegated carpet of many coloured flowers covered all the hills. In the meantime the ice again closed in on the shore and stopped the further passage of the dory. Consequently I had the

company of the Lieutenant and Hansen during the whole
of my stay there. The Eskimo hunted with very good
results, bringing home one day, among other things, no
less than thirteen geese, which they had killed with stones.
On August 11th I had finished all my stations. I said
good-bye to my comrades, who were still ice-bound, and

G. HANSEN AND H. HANSEN ON THE BOAT EXPEDITION.

shaped my course for Ogchoktu. We rowed the ten
miles in about three hours, which was very good work,
as we were heavily laden and somewhat impeded by ice.
On board we had a very hearty welcome, none the less
so as we brought a good supply of reindeer and geese.

It was now quiet in Gjöahavn. All the Eskimo had
left us, and this was the first time for many a month that

Chapter VI.

we were alone. The summer is really not very reliable in these latitudes. On August 16th, rain and sleet set in, and as the thermometer in the cabin only registered $37\frac{2}{5}°$ Fahr., we had to light a fire.

Out of our empty petroleum casks Lund and Hansen had constructed a first-class residence for the dogs. Now they would have a much better home than in the old snow house; hitherto they had been all chained together in the sand, and it was very cheerless for them. Even I myself, after a couple of days, began to get tired of doing nothing, and I decided to carry out a long-meditated plan. As I thought I had noticed some discrepancies in the observations I made along the coast of Boothia Felix during the spring, I determined to estab-lish a station as far north as I could get on the east coast of King William Land. I therefore proposed to estab-lish a depôt there now, for a sledge tour in the autumn, while there was still open water. The boats at my disposal were certainly not specially suited for a fairly long voyage, but by hugging the shore it might be managed. I chose one of our two flat-bottom boats, to which earlier in the summer Lund had fitted a keel. However, the ice conditions were very unfavourable. Ice lay close up to Von Betzold's Point, and to start until it receded was out of the question. Nevertheless, on August 20th, being tired of waiting, I started with Talurnakto as my only companion to try whether, after all, there was any possi-bility of working our way round the point and so getting further on. As there was no wind we rowed; but we did not get on very well. Talurnakto was quite unaccustomed to rowing with two oars, and as he could not keep time, he was always coming into contact with my oars and getting into trouble. Every now and then the flat boat swung round with bow pointing homewards. On board they were following our strange manœuvres and expected to see us back again shortly. Suddenly the secret of keeping time seemed to burst upon Talurnakto, and he rowed liked a good fellow. He rowed bow and I sat in the stern. By that means we travelled along splendidly and were soon at Betzold's Point. The ice, however, was altogether impassable there, as well as farther along the coast towards the east. We therefore landed our cargo, drew the boat up, turned it over, and then returned on foot.

Now followed a long and wearisome time. Every day I wandered out to the point to see how the ice lay. But it was only by August 29th that a land breeze drove the ice away from the shore and we were able to start again. We launched the boat in the channel and put the cargo on board. A fresh breeze blew from the north, with a heavy sea off Schwatka Bay. Before we could set our course for the Duke of the Abruzzi's Point we had to row against the wind some distance along the shore. It was a hard spell of two hours, and we rowed till the perspiration poured off us. Finally we set sail and headed over towards the point. The waves broke constantly over the sides of the boat, and Talurnakto baled like a Trojan. This was Talurnakto's first sailing trip, and he enjoyed it fairly well. We got across the bay before sunset without any further damage than a thorough soaking. We rounded the point and sailed a short distance along the opposite shore towards Abva (Mount Matheson). The first thing we had to do was to get our clothes dry. The treatment of skin clothes necessitates the greatest care, in order to prevent them cracking ; in this the Eskimo are experts, from their life-long experience. The tent was pitched on a little mossy spot on the beach.

Talurnakto handled the " Primus " like an expert. His share of the domestic work, however, had to be restricted to what he could do under my closest supervision, because one never knew what dirty tricks he might get up to. His toilet was grand. Next to his skin he wore a blue woollen guernsey, over this a hunting shirt, and outside an under coat (anorak). His understandings were clothed in a pair of moleskin trousers. All these were worn-out old clothes discarded by Lindström. " I shall darn them during the winter," he said ; but meantime he left the rags as they were. On his head he had an old cycling cap, to which he had attached a dirty collar by way of ornament. Take him all round, he was really a regular " 'Arry," and always cheerful. He smoked and chewed tobacco, and he did all he could to conduct himself like a white man. He took great pride in about six hairs, half an inch long, growing on his upper lip. He spoke with the utmost scorn of men who had no moustache. He was as strong as a bear, and, as he was so willing, he was a splendid fellow to have as help. He had also travelled a good deal, and, among many other places, had been to Eivili (Repulse Bay, an arm of Hudson Bay), and he told me of the many encounters he had had there with the musk ox. The Eskimo are in the habit of goading the animals by shooting arrows at them, and when the animals become enraged they charge the hunters, who kill them with their spears ; in other words, a regular Spanish bullfight.

Before I got on sufficiently intimate terms with Talurnakto to tell him to hold his tongue, he worried the life out of me with his perpetual "singing." I never met any Eskimo really musical. There were, however, slight differences ; most of them could produce four different notes ; Talurnakto, however, contented himself with one—invariably the same, like a bumble bee in a bed-room at night. We were good friends, and he made it a point of honour to copy me in everything he could, thus, for instance, he did his best to use my table cutlery. When he saw that I dried my cup after using it, he did the same, but he used his tongue with which to lick the cup clean, and finally dried it with his shirt.

At 8 o'clock the next morning we started off again ; it was still blowing fresh from the north. After a few hours' rowing we came to a large bay, stretching from east to west. We made repeated attempts to get across, but the waves filled the boat, and we had reluctantly to give up our attempt and go ashore. The next morning we started at half-past four, to profit by the perfectly calm weather. We soon crossed the bay, and were now on the north side of Abva. Here the ice was close up to the land, but we managed to push through; it was so shallow that we had to punt along time after time. The land was barren and waste, with sand and stones, and when at nightfall we had to stop on account of the ice we were unable to find a mossy spot large enough to pitch the tent on. About thirty yards inland, and at an elevation of about fifteen yards above the sea level, I found the skeleton of a whale, and it was also here that I found the first piece of drift-wood I had seen on King William Land.

A disappointing sight presented itself to us the next morning; the whole coast was blocked with ice, and we were consequently compelled to remain where we were. I utilised the time in examining the land. To the north we had a large bay which, according to M'Clintock, must be La Trobe Bay. The whole south strand was sandy and barren and no reindeer moss was to be seen within two miles from the coast-line.

Here we remained for two whole days and nights, and although we continued our journey on the morning of the third day, our delight was but short-lived. We proceeded a couple of miles and reached La Trobe Bay where the ice stopped us altogether. We never moved from that spot till September 16th, but remained there watching the ice. It blew and snowed and we could see nothing in any direction. Evidently summer was over. We took advantage of one of these weary days of waiting to go hunting, and were lucky enough to shoot a doe and her calf. This piece of luck quite turned Talurnakto's head, probably because he had been bored for such a long time. On his way back he began to dance and hop on one leg like a madman. The ice had covered over the small lakes, and in the middle of one of these his heavy body overbalanced itself and he fell down ; the ice was so thin that he went clean through and stuck there. I laughed at him till I was tired, but he freed himself and then performed all kinds of antics to prevent his back freezing.

This boating trip was consequently a failure. It was immaterial whether we laid a depôt here or in Gjöahavn. So we took as much as we could carry, buried the rest under stones, turned the boat upside down and retraced our steps to Ogchoktu. It was twenty-five miles as the crow flies, but taking into consideration all the bends and angles, we must have covered over fifty miles in the boat. It took us three days to reach the far end of Schwatka Bay, but we were often compelled to go somewhat out of our direct course. We pitched our tent there and left it standing as we went on board again the next day. It was our intention later on to go deer hunting in the bay, and we consequently left our equipment behind. The ice had now begun to form, but it was not strong enough yet to bear.

In Ogchoktu everything was in good order. On September 7th the Lieutenant and Hansen returned from their long trip having successfully accomplished their

work. They had reached their goal, Cape Crozier, and
established the depôt there. The narrowest part of
Simpson's Straits had been investigated as well as

WINTER IN GJÖAHAVN.

possible. They found the channel between Eta Island
and King William Land so filled up with shallows, that
it was practically blocked, and the southern one between

Eta and the American mainland was filled up with ice both on their way out and on their return ; but it appeared thick enough to suggest that there must be depth sufficient for our vessel, because if this ice could make its way through the channel, the " Gjöa " ought, with a little care to do the same. This was very important information, as it showed that the North West Passage was not blocked at this point.

It began to look rather doubtful whether there would be much hunting this autumn. No large herds of deer had been seen in Ogchoktu, although there was a solitary deer occasionally. While I was away, Hansen and Lund had made a hunting expedition to the west, in the dory, but they did not meet with much success. Besides, the ice came upon them so suddenly that they were compelled to put the dory ashore and walk home. We had now two boats out on the land. In order to obtain a sufficiency of meat for the winter I determined as soon as the ice permitted to send out hunting expeditions in all directions. Certainly the Eskimo had promised to bring me meat when they returned (and we knew they meant to do so as soon as the ice permitted), but I was unaware at this time how faithful the Eskimo are in keeping their promises, and consequently I did not like to rely on them.

We made a few improvements at the " Magnet," guided by the experience obtained during the previous winter ; the roof was covered with peat, nearly the whole house was buried in sand, and thorough ventilation provided. However, early as it was we could see that winter was setting in. Cold nights, snow-falls, and large flights of migrating birds, were unmistakable signs. The summer had been cold and inclement and there had been very little open water for navigation. We could only hope for better luck next year.

By the night of September 21st everything was frozen over and our second winter had begun.

CHAPTER VII.

THE SECOND WINTER.

THE same day on which the ice really closed up the Eskimo began to make their appearance. The first to come was our friend the "Owl" and his family. He had passed the last few weeks hunting deer and fishing for salmon on the east coast of King William Land at Peel Inlet, together with a number of other families; these latter had been principally occupied in fishing, and had caught a tremendous number of salmon in the river which runs into Peel Inlet. In the rivers here there is really a wonderful quantity of salmon such as is hardly to be found elsewhere. The "Owl" had seen a great number of deer. He had shot twenty of them, of which he let us have the best joints.

The "Owl" told me that towards the south he had encountered an Ogluli Eskimo, by name Tamoktuktu, who lived in an ice-hut with his whole family just below Wiik's Hill. As I had never seen such an ice-hut I went over the next day to inspect it. I found it with the help of the "Owl," who accompanied me. The hut was constructed of eight rectangular blocks of ice, half a foot thick, and about a yard square. They were placed on edge, and bound together with a mixture of snow and ice which was admirably suited for the purpose. The roof was of deerskin. I found the whole family indoors. Puyalu, Tamoktuktu's wife, sat in the background on some skins, fat and contented; round about were lying the bones and remains of fish, and outside the hut was a large quantity of frozen sea-trout. It was too late in the year for drying the fish ("pepchie"), and it could only be preserved in the frozen condition.

Tamoktuktu was just about to start out fishing, and I obtained permission to accompany him. His implements were a "Kakiva" (salmon spear), and a line with bait consisting of reindeer gut, from which was hanging a number of small bright polished pieces of bone and tin. His eldest son came with us and carried a couple of large stones. We started at a good pace for Ristvedt's water, where the fishing was to be. The ice lay on the water bright and clear like a mirror, so that every stone and every substance could be seen below it. Tamoktuktu chose a suitable place for the day, and took the stones from his son, and with these cut a hole in the ice. When he had cut through he cleared the opening with his knife and dropped his line down. The small sea-trout immediately crowded forward to satisfy their curiosity and sniff at the numerous mystic objects on the line, but only to be impaled by the vigilant Tamoktuktu. The "Owl," in the meantime, remained in the hut with Puyalu. They had got to be very good friends, and I could see that the "Owl" wished to tell me something.

Anana.   Onaller.   Kabloka.   Umiktuallu.

NECHILLI ESKIMO PAYING A VISIT.

This was so, because he had promised Puyalu to ask me if she might have the contents of the stomach of the last deer killed. I granted the request, and the " Owl " hurried off to fetch the stomach. A couple of days later Tamoktuktu came with his wife on board to return our visit. They brought with them a great number of fine trout which Lindström immediately bought. As soon as they had been paid they both began to complain how miserably poor they were, and that they had nothing to eat. But we were now too well acquainted with this trick of the Eskimo. We had only too often seen how the Eskimo themselves laughed in their sleeves at us when we had allowed ourselves to be fooled into pitying them and giving them food.

A few days afterwards I took the " Owl " and his family, together with Talurnakto, with me and returned to our tent at the head of Schwatka Bay, hoping to shoot a few reindeer. We were up the next morning as early as 5 o'clock and scrambled to the top of the ridge in front of the tent ; as we reached the crest we perceived twelve deer grazing a little further out on the plain. Happily the wind blew from the animals to us, but the land was very open and rendered it difficult for us to get within range. I let the " Owl " and Talurnakto each go out on one side and I endeavoured to approach from the front. As the " Owl " was the best shot, it was decided that he should open fire. I crept forward like a snake, lying flat on my stomach, and reached a few grass tufts, which gave some protection. I did not dare to proceed any farther, even if I had been able to. The Eskimo are so careless with weapons and I should have got into their firing line. When I had been lying there for some time I looked round for my two companions, when all at once the deer suddenly lay down. I then knew that any attempt to approach was impossible. This period of waiting was a great trial to the patience, and one would certainly require to have more of the hunter's blood in his veins than I to find any pleasure in it. Luckily this time I escaped with only an hour's waiting. Then a large buck raised himself up, and one by one the others followed his example, and began to graze again. I looked for the Eskimo, who were now certainly advancing, but could not see them. Then I heard the report of the " Owl's " gun ; the deer collected together and threw their heads back ; they stood as if they were perplexed, not understanding or being able to discover anything, as the powder emitted but slight smoke, which did not betray the shooter. Then I heard a fresh shot, and the animals began to rush wildly away, straight down towards me ; I lay under cover with my rifle in a splendid position. There was a lake between the animals and myself, and this was clearly their goal. They set out on the ice, still directing their steps towards my hiding place. I should have had to fire directly at the head of the large buck, but suddenly, at about eighty yards from me, the whole flock turned aside and began to run at right angles to their previous course. This was much better for me, as I now had the whole side of the animal to fire at. I fired off all my ten cartridges as quickly as I could. The flock of reindeer rushed in confusion and anxiety in all directions. The result of my shooting was better than I had even hoped for. One animal lay on the ice, another on the bank, and a third limped inland, evidently badly wounded. The " Owl " and Talurnakto now began an eager chase after the wounded deer. I myself had no more ammunition, and as on this occasion I was using a Mauser rifle, I could not use the carbine cartridges of the two others. But now my luck had stimulated my eagerness and I followed on in order to see the end of the hunt. The animal's persistency in getting away on three legs was remarkable. The fourth leg had been hit and he could not use it. We rushed and jumped and finally reached a valley which gave us some protection, and managed to get near to the deer, when Talurnakto put an end to its days.

It had already begun to be less pleasant in the tent. The snow lay on the canvas, and when I warmed the inside the snow melted and made the canvas wet. When the " Primus " stove was extinguished and the heat escaped through the tent, it all froze, and thus formed, as it were, a wall of ice. So I decided to build the tent in. With the assistance of the Eskimo, eight blocks of ice, each half a foot thick, three feet broad, and five feet long, were fetched from the nearest fresh-water pond. We arranged these around the tent and bound them together with ice-mash. We then stretched a deer-skin over to serve as roof and the house was ready. How-

Chapter VII.

ever, it did not come up to my expectations ; it rimed so in the tent that I sat, as it were, in a heavy fall of snow. Consequently, one day my Eskimo built me a real winter " igloo." The " Owl " and Talurnakto lived in a hunting lodge made of ice, lofty and roomy, and all prettily decorated with hunting trophies, deer horns and the like. All building work was carried out whenever the weather prevented hunting.

One day, tired of this fruitless wandering around, I left my two companions and jogged home to my " igloo." In our camp I found Anana and Kabloka engaged in drying skins and passing the time according to their custom. I went on to my hut and made coffee and invited the women in to gossip for a time. They always highly prized an invitation to come to see me, but they never came when the men were at home. We chatted comfortably for a good hour, when they went back to their work. But towards evening both the women began to be very anxious for their husbands, who had not yet returned home. Nightfall came on, and Anana came again and again, much worried, to me, to ask about her son, and I comforted her as best I could. I lit as many lights as I could within the hut, so as to render it possible to be seen by the two hunters in case they might have lost their way, and then I joined my two female friends. Poor things! they sat there huddled up and shivering, very frightened. This is the worst time for the Eskimo, as they have not yet collected fat for light and warmth. As I sat and chatted with them, old Anana was suddenly

seized with a fit, at least, so it appeared to me. Her
previously sallow yellow complexion turned an ashy grey,
her lips trembled, her teeth chattered, and she uttered

THE "OWL'S" ICE PALACE ON KING WILLIAM LAND.

the most unearthly inarticulate screams. I took hold of
her and began to shake her as hard as I' could, so as to
bring her to herself again. Then Kabloka came to me,
laid her hands on my arm, and whispered solemnly,

Chapter VII.

"Anana, angatkukki angi!" which interpreted means, " Anana is a great witch ; let her alone." " Nonsense," I replied, in good Norwegian, and I continued to shake the old crone until I had shaken the witchery out of her. I then went over to my hut and fetched a light as well

Talurnakto.   The " Owl."   Ristvedt.
A HUNTING PARTY.

as my good " Primus " stove.   Warmth and light proved to possess the necessary magic power, and both the women soon had what they required, their good humour returned, and the hut echoed with song and cheerful conversation.   At 9 o'clock the two missing ones returned.

They had gone a good deal out of their way, but the light from my hut, which they had observed a long distance away, had put them on the right path again.

On October 2nd I returned again to the " Gjöa," and let Ristvedt take my place. On my way back I encountered the tracks of a she bear with two young ones that had evidently only just been born ; they were making their way towards the south to milder regions. This was the first track of bear we had seen in the neighbourhood of Ogchoktu. When I came on board, Umiktuallu, or " the murderer," as we generally called him after the dreadful event, had arrived from the American mainland. He had come over from Navyato—" Hunger Bay "—but the ice was not very good for travelling. He had killed thirty-five deer from the kayak, and it was his intention to bring us the joints as soon as the ice was fit. This was a splendid addition to our relatively poor stock. Umiktuallu also informed me that large herds of deer passed over the ice past Todd's Islands. A piece of news that excited our curiosity to the utmost was that Umiktuallu had met a Kilnermium Eskimo, who had informed him that a " kabluna "—white man—had visited the Kilnermium race (which name was applied to those who lived near the Coppermine River) in the month of March, together with an Eskimo family who lived a long distance away. Later this report proved to have been correct. We now determined to try our luck at hunting towards the west, where Umiktuallu had seen so many deer. Lund and Hansen, therefore, travelled

together with him towards the west, with sledge, five
dogs, and provisions for fourteen days.

I got a letter from Ristvedt, brought to me by an
Eskimo courier, saying that he had moved the tent up
near to the "most northerly north hill," where he had
seen tracks of large herds of deer. There were certainly enough deer this year ; they had found the ice
conditions around Eta Island favourable for them, and
therefore wandered around us. Contrary to what was
the case in the year 1903, in 1904 the deer were well
nourished and fat, although, perhaps, not so fat as the
deer on Spitsbergen. It is really phenomenal what an
amount of fat—a layer of several inches—the Spitsbergen
deer, so thin and lean in spring, can put on in the
summer. The ptarmigan also had now begun to migrate
towards the south. Large numbers of them passed us,
and we secured as many as we could. October 15th
was a lively day in the harbour, inasmuch as our hunters
and four Eskimo made their appearance together. Lund
and Hansen had killed nine deer, which they had walled
up in a snow hut in order to bring them on board
later. However, we put off fetching them for some
time, and when, finally, in the winter, we sent to fetch
the meat, we found it had been stolen by some Ogluli
Eskimo. Now the whole stock of meat obtained by
ourselves consisted of twenty deer, but we had no fish
at all ; however, the Eskimo, brought us in plentiful
supplies of both for the winter. On the whole, I
must give these Eskimo an excellent testimonial for

reliability ; without any exception, they always fulfilled their promise to bring us meat and fish.

This winter we discovered a new method of providing water for our fire brigade, and as possibly some of my readers may happen to need such a fire station, I shall explain our arrangement. Every first day in the month the chief of our fire brigade, Lund, went out with an ice-bore and measured the thickness of the ice. At the side of the vessel he had built a large roomy igloo, and on that day he cut his way down in the ice within the igloo, but he never cut right through, always leaving a thin layer of ice ; thus, if there were six feet of ice, Lund only let the fire tank have a depth of 5 feet 10 inches. Now, if the fire alarm were given, the chief of the fire brigade had only to spring down into the tank, seize hold of the ice-bore, which was always ready there, and immediately bore through the ice, when the water would rise up through the hole, and very quickly fill the large well. Opinion was considerably divided as to the reliability of this arrangement, particularly as to whether the water would fill the well quickly enough. I rather shared this doubt myself, and even the chief of the fire brigade did not place absolute faith in the arrangement. When the ice was thickest, therefore, we tried the experiment, and sounded the fire alarm. Lund sprang unconcernedly into the well, seized hold of the ice-bore, and began boring ; but this time he came up pretty quickly. The pressure from below was terrific, and the water foamed up into the well, so that the chief had to be thankful that he got nothing worse than a wet skin. By this method he had only to cut the ice once a month, instead of having to do so every morning.

The return of the Eskimo again imparted a lively and variegated aspect to our little harbour. They came on board, as a rule, generally of an evening in great crowds to visit us or to introduce new friends. They were always gay and happy, and we became very good friends with them. It has always been believed that the air in the Polar regions is absolutely pure and free from bacilli ; this, however, is, to say the least, doubtful, in any case as far as the regions around King William Land are concerned, for here the Eskimo nearly every winter were visited with quite an epidemic of colds. Some of them had such violent attacks that I was even afraid of inflammation of the lungs, and as nearly every one of them contracted the illness, it must in all probability have been occasioned by infection. Happily those on board the " Gjöa " escaped, but we certainly took due precautions. We had great trouble to put a stop to the spitting habit. The Eskimo are very bad in this respect, but when we had them some time under treatment they improved and paid more attention to our prohibition.

As I both desired to see the large Eskimo camp which was said to exist on Navyato and to barter for a supply of fish I took Helmer Hansen with me to visit it. We started on October 23rd, accompanied by three Eskimo families who were going the same way. For a change we went on snow shoes. The snow had not yet packed together, and as the temperature was not more than 13° below zero it was very suitable for snow-shoeing. At

AKLA AND SON.

half-past four in the afternoon we reached the top of Elling Hill, fifteen miles from the station ; here we found four igloos, which we baptised " Hotel Elling Hill." Together with my friend Poieta and his wife we took possession of one of the igloos, and thanks to Nalungia's womanly assistance we were soon snug and cosy. After finishing our meal and conversing a little we lay down in our hammocks. Naturally the little boy cried every now and again and had to be quieted with kisses and nursing. I grinned at it, but Hansen was evidently delighted ; I perceived that this nightly scene sent his thoughts far, far away where *his* Nalungia was perhaps now kissing and singing to another little fellow.

The next morning we set out on Simpson's Strait in a southerly direction. The irregularities of the ice were not very great, but they nevertheless bothered us considerably after having been accustomed to the level snow fields. At 4 o'clock in the afternoon we reached Navyato, where to our surprise we did not find altogether more than ten huts ; considerably less than we had expected. We were, however, very well received by a number of our old acquaintances who were here cod-fishing. Navyato is the shore surrounding a tolerably large lake, a few miles to the south of Point Richardson, or, as the Eskimo call it, Novo Terro. Navyato· is situated not far from the bottom of Hunger Bay, which latter owes its name to a large number of skeletons of Franklin's men having been found there, plainly having starved to death on their way to the south. It is an irony of fate that this sinister name has been applied to what is in reality one of the most beautiful and lovely spots on the American north coast. In spring, when the channels are opened, enormous quantities of large fat

AN ESKIMO CAMP IN WINTER TIME.

[*Photograph taken by Moonlight.*]

salmon are met with. A little later the reindeer arrived in countless hordes and remained here throughout the summer, then in the autumn an unlimited quantity of cod can be caught, and yet here—in this Arctic Eden—those brave travellers died of hunger. The truth probably is that they had arrived there when the low land was covered over with snow ; overcome by exertions, worn out with sickness, they must have stopped here and seen for miles before them the same disheartening snow-bedecked lowland, where there was no sign indicating the existence of any life, much less riches, where not a living soul met them to cheer them up or give them encouragement and help. Probably there is not another place in the world so abandoned and bare as this is in winter. There when summer comes and millions of flowers brighten up the fields : there where all the waters gleam and all the ponds sing and bubble during the short freedom from the yoke of ice : there where the birds swarm and brood with a thousand glad notes and the first buck stretches his head over the ice harbour : there a heap of bleached skeletons marks the spot where the remains of Franklin's brave crowd drew their last breath in the last act of that sad tragedy.

On this spot, which conceals so many sad mementoes, the Eskimo live gay and happy until darkness comes on and throws its iron cloak over these regions and all within it. They caught large quantities of cod in three to four fathoms of water, using a line of reindeer gut and as hook a bent nail. The same afternoon that I arrived

Chapter VII.

I was presented with live cod which were at once put into our pot, and they were wonderfully tasty. After our long and tiring march we sat down and enjoyed our fresh fish, imagining ourselves home amongst the rocks of Norway on a summer's evening. We had, however, come here less for eating than with the view of buying cod, so in the succeeding days we got to business ; I tried my luck at fishing out on the white ice, although without any great sucess. The days we passed at Navyato were cold and raw. The temperature kept at about 13° below zero Fahr., and nearly every day was showery and windy. The cold in autumn and spring is what one feels the most ; although, of course, the thermometer is considerably lower in winter one feels it less, probably because one is better clothed and better prepared for it.

The Eskimo were out fishing from morning till nightfall, though as they are by no means early birds they did not begin exceptionally soon. They ate very little of what they fished, the bulk being placed in "perura" (depôt) ; thus these people made some provision for the future, in any case they laid in sufficient stock for the first portion of the winter when it is so difficult to get food. We were otherwise very comfortable in Navyato. Poieta went off to fetch some buried reindeer meat, but he allowed Nalungia to remain behind ; she took care of our things, and did the best she could for us. Poor Nalungia ! I perceived that in the beginning she was rather afraid of being left alone with the two white men, but when she perceived that we neither attacked her or her baby, she was very soon happy and contented. We hadn't much for her to do, but still she was useful for cleaning the fish. In the beginning also her little boy Aleingarlu regarded us with considerable suspicion. He cried and screamed at all our attempts to curry favour with him. But gradually he gave way, and after a short time we had to be continually rubbing noses with him, this being the Eskimo manner of embracing.

After the lapse of a few days we left Navyato with our sledge fully loaded with fish. Many of the Eskimo accompanied us on our return trip, amongst them being our oldest and best friend, Teraiu. He had not forgotten our trip to Kaa-aak-ka, and was impudent enough to continually remind me how he had then showed himself to be a miracle-monger—when he intentionally led us astray. The old fellow had now contracted a severe cold and accompanied us in order to get some medicine. During the trip he had one attack after the other, and coughed up a considerable quantity of blood. Old Auva, the liveliest of them all, we left in a very wretched condition, suffering from some stomach trouble ; when we started she was still able to sit by the fireside, but she died a few days afterwards. We passed the first night on one of Todd's Islands in an old snow hut. On these islands a few skeletons and other traces of Franklin's expedition had also been found. Teraiu told me that he had heard tell about all the white men who had lost their lives there.

Chapter VII.

He showed me a large white stone on the island on which we were passing the night that had been set up in memory of the dead. This island is called Keuna by the Eskimo. Our arrival with a large sledge load of fresh frozen cod naturally occasioned much pleasure to those on board.

In accordance with the experience gained in the previous winter, we continued our improvements to the "Gjöa." The winter roof over the boat was better fitted up, and we arranged an inside door so that we had, as it were, a second house. This arrangement had also the advantage that we were better cut off from the Eskimo, which at times was really necessary. Of an evening Lieutenant Hansen locked the door, and there we all sat safely and securely as in a fortress. We placed our little American steam bath on the floor. The Lieutenant and I made constant use of this steam bath in the course of the winter; it worked splendidly, and, living as we were in such close quarters, heavily clothed, etc., it was really indispensable to us. We had no dressing-room to spare, but we used an old butter cask turned upside down. What was by no means pleasant was the cold douche which we always obtained from the rime formed on the roof from the frozen steam. Lieutenant Hansen arranged this bath splendidly, and also installed the electric light—three complete lamps. Indeed, he went so far as to install the electric light on deck; thus one sat quite comfortably in the cabin, pressed a button and—yes, we ought to have got light, but there was no light. This was the only fault of Hansen's electric light installation, it never gave any light; we had just to take our bath in the darkness. It should, however, not be concluded that the Lieutenant was a bad electrician. The facts of the case are that it is not given even to the best of mortal electricians to "make something out of nothing, or out of unsuitable materials" and say "let there be light"; although we were often very proud of what we managed to do with very unsuitable materials.

One question that the Lieutenant and I had often discussed was how we could protect ourselves against the Eskimo, in case that they should take it into their heads to do anything. There was now a great number of them collected around us and if, for example, they were not very successful in their catches, our provision tent was exposed to them. We had, therefore, to teach them to regard us and ours with the greatest respect, and at last we hit upon a method of accomplishing this. A powerful mine was buried beneath a snow hut at a good distance from the ship, and a train laid from the ship and well covered with snow. When that was ready, we collected the Eskimo together on board. I spoke to them about the white man's power; that we could spread destruction around us, and even at a great distance accomplish the most extraordinary things. It was, consequently, for them to behave themselves properly and not to expose themselves to our terrible anger. If they should play any tricks on land, for example, over there

by the snow huts, then we should merely sit quietly
on board and do so . . . . . With a terrific report the
igloo blew up, and clouds of snow burst high into the air.
This was all that was required.

Silla was out again and had puppies. This time she
treated them with greater care than before, and consequently managed to keep them. When they were three
weeks old we took them from the mother and placed
them in a locker on board. It was really touching to see
Silla, who was placed out on the ice. We had a ladder
outside the ship which we ourselves found considerable
difficulty in mounting, but the unhappy mother climbed
up and found her young ones again. Nothing kept her
back; she would even have gone through fire to have
joined her offspring.

"Uranienborg," our astronomical observatory, was also
considerably altered and improved. More particularly
the assistant, Ristvedt, here, as everywhere else, knew
how to make himself comfortable. The establishment
consisted of a half igloo and a whole igloo. In the half
igloo stood the observer, Lieutenant Hansen, in the
whole igloo lay the assistant. It was a topsy-turvy
arrangement, but it was actually the case. We must,
however, admit that the good Ristvedt was right. Why
should he stand outside on a cold winter's night and
freeze when he could do his work just as well lying down
in a comfortable room? I visited them now and again
whilst they were working. I first met the Lieutenant
standing outside in the half igloo with the instrument to

SILLA AND HER PUPS AT THE STORE SHED. GJÖAHAVN.

his eye observing the movements of the heavenly bodies. These are slow in their movements, but on such a winter's night they seem slower than ever for him who has to be on the constant look-out and catch them to the second. It was the very deuce! the Lieutenant had to take his mittens off and rub his white frozen hands, stamp his feet, and look again, to find the star gliding into his ken. " Now!" he called out in the stillness of the night. A distant echo answered him from the deep; I followed the sound and went to pay a visit to the echo. Passing through a little opening into the other igloo I suddenly stood in a light, warm, cosy room, and there lay the assistant in his sleeping-sack, with a large oil lamp in front of him; his pipe lighted up his features, and the only thing missing was a toddy glass steaming on the table. Yes, he knew how to make himself comfortable.

On Sunday, November 20th, as we sat at breakfast, we were surprised by the visit of an Eskimo: a perfect stranger to us. The manner in which he entered showed that he had been among " people "; his clothes were also quite different from the Nechilli race. Our astonishment was not less when the fellow addressed us in, if not perfect, at least very intelligible English. " Give me 'moke!" We set pipe and tobacco before him and he filled his pipe like a perfect gentleman; he then introduced himself: " I am Mr. Atangala." This began to be very interesting, I sat and observed him, eager to see what his next step would be; I was, however, soon brought to myself and reminded that it was my turn to

Chapter VII.

take the next step. "Might I ask, sir, what is your name?" I blushed at my ill-manners, bowed slightly, and gave my name. The introduction being over, he was evidently more at ease. He intimated to me that his family were on deck, and I immediately remedied my previous behaviour by inviting his family down. They did not need to be urged. The woman was a tall, dark, typical Eskimo; her name was Kokko, and she was about thirty years old. Their son was about ten years old, and at first sight one saw that he was an exceptionally unmannerly boy. Atangala told me that he and his family had accompanied three white men from Chesterfield Inlet, in Hudson's Bay, to Coppermine River. This confirmed the report of the "murderer," Umiktuallu, about six weeks ago. From Coppermine River he turned his steps homeward, where he learnt that a vessel was lying in Ogchoktu—a trifle of about a couple of hundred miles away—so he determined to visit us and see if there was any business to be done with us. He boasted a good deal about being able to write, and at his request we brought him pencil and paper, but his skill in this useful art was not very remarkable. After a great deal of trouble and time he managed to sign his name. A few years ago he had accompanied an American whale fisher overland from Hudson's Bay to Winnipeg, and during his stay there he had become acquainted with all the latest discoveries such as the telephone, railways, electric light, and— whisky. He was especially interested in the last, and

Assistant's Compartment.　　　　　　　Observer's Compartment.

URANIENBORG.　THE ASTRONOMICAL OBSERVATORY, IN WINTER TIME.

asked eagerly about it. I tried to explain to him that the teetotal movement was the latest advance in the region, but he would not listen to it. At last he asked straight out for some spirits, but we did not give him any. It was of the greatest interest to us to learn that two large vessels were at Katiktli (Cape Fullerton). I immediately began to think of getting into postal communication with the outer world by these two ships, and I asked Mr. Atangala if he would be willing to act as post-boy. He appeared willing to accept this position. The boy, however, was an awful nuisance. At one moment this ten-year-old hobbledehoy lay on his mother's breast and took a drop of milk and then snatched the pipe from his father's mouth and began smoking.

The day after this great event Lindström got sick. It was a stomach trouble such as he had had the year before, although not so bad. I doctored him with hot plates and opium, which seemed to relieve him considerably. In the meantime we had to have another cook, so Ristvedt and Hansen did the work in turn. These gentlemen surprised us by supplying dishes which rivalled those of Lindström ; indeed, in many respects even—but no! Lindström might be annoyed if he saw these lines, so I shall pass to another subject. One of these days Umiktuallu and Talurnakto came with a sledge full of deer joints and fish ; so, thanks to the Eskimo's faithful assistance, we now had full winter supplies.

Everyone was very busy writing letters ; the post was going to start in a few days, and we all had to have our own correspondence ready. Atangala was a constant guest on board and allowed to make himself thoroughly comfortable. His dogs, however, were in a miserable condition. They looked like thin wolves, sneaking around and searching for prey. I felt sorry for the animals, but what were we to do? We hadn't enough meat for our own dogs, let alone for those of others. Our dogs' pemmican we had thrown overboard, as this food was not liked by our own dogs; we therefore gave a little of it to the poor post-dogs, but they were worse off than ever, because the pemmican acts like a violent purgative. On November 28th the post sledge was ready, and, in spite of breeze and thick mist, the letter carriers drove off at 11 o'clock in the morning. I thought it advisable to let Talurnakto accompany Atangala. I was unacquainted with Atangala, and, for all I knew, he might be the greatest rogue in the world. It was very comical to see Talurnakto swelling with pride at the important and serious commission entrusted to him; he carried the letters which were addressed to the ships at Cape Fullerton in a bag strapped over his shoulders. I saw Atangala smile, but it was only later on that I knew why. So they started and disappeared in a snow-cloud past Fram Point.

One of the essential conditions for the success of Arctic exploration, for working together through thick and thin, is for every man to have all his time occupied. It is the leader's duty to see that such is the case, and for long periods this is by no means an easy task. But

The Second Winter.

idleness has a very demoralising effect, and for this reason alone it is not advisable to have too many people with one; work can always be found for a few, but it is almost impossible to find constant work for a large number. This important duty was, however, not very difficult for me, as my comrades always met me half-way, so that if I could not think of anything for them to do they had always some proposal to make. Lund was famed as our great inventor in all matters. If one took an idea to him it was not long before it was carried out, although when any smith's work was involved he had to get the help of Ristvedt; but when blacksmith Ristvedt and architect Lund worked together, then nothing was impossible. We had brought with us from Godhavn a quantity of fire-proof bricks, which we had intended to use for bricking in our petroleum stoves so as to keep the heat longer. For some time I had been considering how these bricks might best be utilised, and I went and consulted Lund on the matter. We discussed the problem for some time and at last agreed that we might just as well construct a new stove as brick in the old one, and Lund undertook its construction. On the same day as the post left, he surprised me by showing me the stove all completed. He had taken one of our large metal boxes and lined it with bricks and then fitted a door at one end. His intention was to set one of our " Primus " stoves, which gave out a powerful heat, with the burner through the door and thus heat the stove, the bricks of which would have retained the heat long after we had

Chapter VII.

put out the " Primus." Apparently it was an excellent scheme and the Lieutenant and myself were delighted to get the apparatus fixed up in our cabin and took the deepest interest in its installation. Everything being in order, the " Primus " stove was lighted and set in place. Lieutenant Hansen was very much occupied with a number of calculations and I was busy with my own affairs. The " Primus " stove had not been burning long before I perceived a peculiar sharp penetrating smell. I looked at the Lieutenant to see if he had perceived anything, but he was deeply buried in his work. I did not say anything but merely stood up and betook myself to the observatory, leaving the further testing of the stove to the Lieutenant. I, for my part, had had enough of it —the smell was simply awful. " This is awful " suddenly exclaimed the Lieutenant, starting up and throwing his pencil down ; " What the deuce is making such a stink here." I was quite out of the door leaving the Lieutenant inside in a thick suffocating atmosphere, whose odour reminded one of all the dogs of Godhavn and " Gjóa." The truth was that the bricks had been lying about unprotected on the dogs' favourite exercising ground and one can imagine the result when exposed to the influence of our wonderful " Primus " stove. I was already some distance up the hill towards the observatory when I heard a great noise on board, and turning round to ascertain the cause, I perceived Lund's ingenious stove had been thrown out. There only remained behind one souvenir—the smell which had penetrated into the walls of the cabin—and it was some time before we could get rid of it.

" Cherchez la femme." Even here in this deserted spot one must now and again utter this cry of perplexity or, more correctly speaking, this lamentation for man's weakness and frailty. Umiktuallu, who was again fishing at Navyato came to us to borrow our dogs for a couple of days, and told us that Talurnakto, when in Navyato, had made over the letters to Atangala and had gone off in a southerly direction with one of the Eskimo women as husband number two. I would remark that this was the only case of breach of trust that I ever experienced among the Nechilli Eskimo. Umiktuallu further told us that a young woman we knew had died in child-birth. This was the second death this winter.

On December 14th our expedition had a new companion, as I adopted a little ten-year-old boy called Kaumallo. The poor little fellow had no parents, and was left to take care of himself; it was a wretched sight to see him wandering about in his rags. He suffered from rheumatism, and had some difficulty in walking; he also suffered from another serious physical defect. We were all glad to get him on board, and arranged as best we could for him. The first thing we did with him was to send him up to the " Magnet " to undergo a thorough cleansing. Wiik cut off his long hair and washed and scrubbed him so that it was really a wonder that he escaped with his life, but he came out of it all right, and was clothed afresh from head to foot. Lund

Chapter VII.

arranged a bunk for him on one of the benches in the cabin, and altogether made him very comfortable. However, his participation in the "Gjöa" Expedition was not of long duration—he could not stand the good living. The change in food upset his stomach, and we consequently had to diet him. While we took our ordinary meals, Kaumallo's diet was suddenly changed to oatmeal porridge, and this offended him so deeply that at last he refused to eat. Under these circumstances, the only thing to be done was to send him off again. The next day he was just as dirty and disgusting as before, but he was evidently thriving.

Now the Eskimo streamed in large crowds to our harbour. They had finished their fishing in Navyato, and other places; and we desired as far as we could to pass Christmas quietly at Ogchoktu. There was but little doubt that our presence imparted a special attraction to this spot. In any case a great number of them began to come and beg for food, although they had more than enough for them and theirs for the winter. Some had little enough, partly owing to bad luck and partly to idleness. It was by no means pleasant to be surrounded by all these beggars; we had to resolutely refuse to give them anything, because it was absolutely impossible for us to support such a great number of people. There was no help for it, they must go somewhere else where food was to be found. This winter most of the Eskimo built their huts in Saeland Valley, and the encampment had quite an imposing appearance.

The Second Winter.

Now we had the darkest time of the year, the sun being below the horizon the whole of the day. We had to use artificial light all day long. With a view to the long period of darkness, I had provided the Expedition with extra patent lamps in which the petroleum is heated. and, as it burns like gas, it gave a very powerful light. Even during the first winter these lamps were constantly out of order and caused a deal of trouble, but this year they absolutely refused to work at all. The lamp department was under Lindström's management, and he was quite in despair, although he made persistent efforts to remedy their defects. But at last he had to give in, and we were compelled to condemn our whole stock of lamps. My great mistake was that I had blindly relied on the patent, and had not taken with me any ordinary lamps, so that towards the end of the winter the lamp department on board the " Gjöa " was at a rather low standpoint. One photographic lamp, one compass lamp, and two lanterns were all that we had. With respect to lamps, Lund, of course, kept on making discoveries, each an improvement on the previous one. At an exhibition they would all have been taking prizes if they would only have burnt, but this was where they failed.

Christmas was approaching and everyone was making preparation for it. As in the previous year, most of the work again fell on Lindström's shoulders, although Wiik also had his hands full. Christmas Eve was very drizzly, though this was of but little importance as we could pass the evening comfortably indoors. Flags were floating all over the ship and on land; even the igloos of the Eskimo were decorated with the Norwegian flag. We had a splendid banquet, although, perhaps, not so fine as in the previous year. Thanks to Wiik's remarkable activity there were, if possible, still more presents than in the previous year. At the previous distribution it so happened that Lindström had received scarcely anything but watch-stands, which of course he could make no use of. Many of the others had also received watch-stands and, like Lindström, did not know what to do with them. Now Wiik went round and collected all watch-stands which did not actually belong to Lindström. He made different sized packets of them and addressed every packet to Lindström with greetings from various known and unknown persons at home. Lindström's excitement on receiving each packet was as great as his disappointment when on opening it, the everlasting watch-stand stared him in the face. With such like tricks and games our little circle spent a happy evening, and our second Christmas, with its gladness and its sadness, passed to everyone's satisfaction.

The Eskimo also kept this time as a holiday. Umiktuallu was an expert in the construction of igloos. After a great deal of work he succeeded in joining up three neighbouring huts to form one, thus constituting quite a banqueting hall, where the Eskimo passed the evenings in dancing, singing, and gymnastic exercises. Altogether Umiktuallu was a very capable fellow. It was he who

SNOW HUTS IN THE LINDSTRÖM VALLEY, NEAR GJÖAHAVN.

had made the largest catches this year. It is a custom at Christmas time to cart home from the depôts whatever had been caught. This cartage is not effected by the sealers themselves, but naturally by the others who help to eat it. Good catches had been made by the Eskimo, who in the course of the summer were encamped towards the north, and as they had consequently a good deal to bring in, they often borrowed our dogs.

They first began seal fishing in the middle of January, and this was by no means too soon, because the stock of provisions had got down very low. I am inclined to believe that it was superstition which prevented them beginning earlier. As far as I could see, the moon had to have a certain position before they dared go seal catching. They look upon the moon as an important sacred body, according to which they divide their time. Their superstition often stood in their way. Just at this time there occurred something rather characteristic. A great number of the women folk had constant sewing work from us, for which we made them some slight payment, much to their satisfaction ; but one fine day they all declared that they could not work. We questioned them and learnt that the first seal had been caught, and that the women had eaten its flesh ; it was consequently impossible for them to do any work other than their own before the sun was at a certain height in the sky. We explained to them the stupidity of the whole thing ; we promised them higher wages, indeed almost begged and entreated them to continue their work for us,

Chapter VII.

but all in vain; they were proof against all ordinary human arguments. Here God or the Devil stood behind and the women refused to comply with our request, although otherwise they were very submissive and always ready to oblige. Old Navya was the only one who had been sensible enough to abstain from eating seal flesh. She now came on board and sewed for us from morning till evening, the object of all our admiration.

Old Navya was a widow, with two children : the one a daughter about twenty, and the other a fine darling boy of about twelve to thirteen. The name of the daughter was Magito, a nice little woman, who had set many a white man's heart on fire ; she was married to Kirnir, a brute of a fellow, who would sell her for a rusty nail. The relations between Navya and her son were really very touching. He followed her wherever she went, treating her with a respect worthy of imitation, and she, for her part, was very fond of the boy, so that it really formed a very pleasant sight. By and bye we learnt that these two had passed through a dreadful experience, which, even had they not been mother and son, would have sufficed to bind two persons together. Some years ago, Navya had been with her husband to fish far away inland, and had taken the boy with her. The fishing was very bad and her husband got sick, so that Navya had to endeavour to get the food herself ; but the fish went off and her husband got worse. At last the man died, and it was impossible to catch any fish, and as there

MAGITO. THE BELLE OF OGCHOKTU.

was no other food to be had, both the mother and child
were forced to partake of portions of the body of their
husband and father. This dreadful tragedy seemed to
cast a cloak of melancholy over old Navya. She was
naturally very bright and gay, but at times had fits of the
deepest melancholy, during which she tenaciously clung
to her boy.

On January 11th I took the census in Gjöa Harbour
and surroundings; I found that there were altogether
sixty souls, distributed among eighteen families. One
day, about the middle of January, Lindström came and
told me that some tins of preserved meat had been stolen
from our store on deck. We had taken on board from
the provision cellars the preserved meat we needed for
the winter and placed it on deck, where it was dry.
From the first I had thought that this might offer some
temptation to the Eskimo, especially at this time, as they
had hardly any food; indeed, I was surprised it did not
happen long before. Eskimo were in the habit of coming
to the ship daily, and it was very easy to snatch up a tin
when passing and conceal it under their clothing. As a
few tins had been taken in this way, I sent for the "Owl"
and Umiktuallu, and told them that the Eskimo had been
pilfering from us, and that I must be informed who were
the guilty ones. They regarded the matter very seriously,
and it was plain that they considered themselves respon-
sible for the behaviour of their kinsmen. They left me,
and a couple of hours later returned and mentioned four
or five as the guilty ones. They were all Ogluli Eskimo;

none of the Nechilli tribe had offended in this respect. I had the guilty ones brought on board before me, and amongst them found my great wizard, Teraiu, as well as his brother, Tamoktuktu. The offence was limited to each having taken away one tin. I spoke very severely to them, and forbade them from that time ever to come on board again, whereupon they slunk away very sheepishly.

This winter I had two magnetic observatories, one constructed by Wiik in the summer, and consisting of old sails and tarpaulins. When winter came on we covered it over with snow, thus making a splendid winter observatory. One of our travelling instruments was fitted up here. About a hundred yards to the north I had a large igloo built, thick ice windows being let into all the walls, so that I had as much light as possible the whole day long. The instrument here stood on a snow socket, and it was so solid that no movement of the instrument, which weighed 66 lbs., was observed during the whole of the spring. As it was of importance to make simultaneous observations from time to time, I arranged it so that we could call from one hut to the other. When spring approached, it seemed as if the magnetic work would be more than Wiik and I could carry out alone. Lieutenant Hansen therefore offered his services, which I accepted thankfully. He had first, however, to go through a course in the manipulation of the instruments; but, as he had been used to handling his own instrument, that did not take very long, and after a couple of weeks he was able to take up his position as assistant.

Before the end of January the Eskimo had more luck with their sealing, which at first had been so unfortunate. On February 1st they began to make arrangements for moving, and brought their provisions out on to the ice. A few days later they decamped and we were considerably relieved to be rid of them; the constant begging for food had been a great worry to us. We now began to fit out the sledging expedition which had been planned long ago for the approaching spring. Lieutenant Hansen and Ristvedt were to endeavour to reach the east coast of Victoria Land and to map it; this is the only portion of the North American Archipelago which has not been mapped. As has previously been stated, the depôt for this expedition had been placed on Cape Crozier, about a hundred miles from Gjöa Harbour. The first question was that of the dogs. We had an excellent team but this was not enough, consequently the Lieutenant and Ristvedt went on a trip to the Eskimo who were about twenty-five miles away from us on the ice, and two days after returned with four large dogs they had obtained in exchange for a few iron bars. These dogs, however, were half starved and had to be fed up before they could be used. Then there was the next question—*food*, which is almost as important as the dogs themselves. All our dogs' meat had disappeared, and all that remained was some pemmican for human consumption;

Chapter VII.

this would not go very far. Our pemmican consisted of 50 per cent. beef fat and 50 per cent. horse flesh, which was dried and crushed. These two substances were melted together and made into bricks of about a pound weight, thus facilitating the packing and the transport. It was originally the Indians who taught us the use of this food, so that all Arctic explorers owe them a deep debt of gratitude. Pemmican has a very excellent flavour, occupies but little space, and can be eaten raw, roast, or boiled, and as provision for a sledge expedition it is indispensable. We had, further, with us a number of boxes of dog's fat which came from the second "Fram" Expedition, and was in very fair condition. We had also a quantity of oats and oatmeal, as yet untouched, none of us caring very much for it. All these things were handed over to Lieutenant Hansen, so that he might experiment with them. The dogs seemed to be very fond of dog fat mixed with oatmeal and thrived very well on it, so Lieutenant Hansen made up pound rations of these foods mixed together. This constituted a dog's daily portion. The manufacture went on on a large scale, the whole ship being a patent dog's food factory, under the management of Godfred Hansen. The work proceeded very rapidly and soon the whole question was settled. Then the remaining implements were examined and Lund was fully occupied working on the sledges, which had been used a good deal from time to time and required fitting with new runners.

Whilst these repairs were going on a message came

The Second Winter.

that Talurnakto, the missing postman, had come to Navyato and begged to be forgiven. We had missed him considerably both on account of his ability as a workman and also by reason of his constant good humour, and after all his offence was not greater than that of others who had run away with other men's wives; indeed it was even less, as in this case he took the whole family, husband and children, with him. Consequently we sent him a message that we would forgive him and replace him in his former position as handyman on board. One evening in February we were told that Talurnakto had arrived. This time he was afraid to come down directly, so I sent to fetch him. His appearance had undergone a sad change since the last time we saw him. He had plainly had a very trying task as a lover. His round glad face was thin and drawn out and bore the stamp of the deepest despondency. He had no cherished mementoes of his amorous adventure. All his property, knife, spear, and pipe, had been given over to his beloved, or rather to the husband, who required some recompense for his liberality. Poor Talurnakto returned from his escapade stripped of almost everything.

On February 7th the Lieutenant made the first sledge expedition of the year, driving to Kaa-aak-ka in order to take a number of magnetic observations. This was rather early for commencing any out-door work, but as we had a good deal to do we had to bestir ourselves, and it was intensely cold. It is remarkable how everything becomes

Chapter VII.

brittle and friable in such cold. One day Lund was working at the runners for the sledges and the piece of hickory he was working was lying over a couple of boxes two feet high, when by an unlucky movement he let the piece fall on the deck where it broke into fragments. Hickory is undoubtedly the toughest wood in the world and ours was specially procured from Pensacola, but in this cold region, green-ash, for example, is much to be preferred. One Saturday afternoon we were surprised to see our sledge harnessed to five dogs come tearing in to the Bay and swing sharply up beside the ship, without any driver. A short time after information was brought to us from the " Magnet " that a dark spot was visible on the ice. Then Talurnakto came up steaming and puffing. He had been thrown out of the sledge and the dogs had run home. The " Owl " came one day occompanied by another Eskimo and told us that some Ogluli Eskimo had been stealing from our provision tent ; they brought with them an unopened tin of butter they had seized. On examination it was found that a quarter box of sledge bread, ten bricks of pemmican and the above-mentioned tin of butter were missing. If we had not been informed, probably we should never have noticed the robbery, as the thieves had left everything in perfect order. I can only say they were easily satisfied. It was old Teraiu, together with his brother Tamoktutu, who were at the head of this expedition, which was carried out early one morning. As reward for his honesty, the "Owl" was given a large hatchet and the other, a knife, As they went off I told them to greet the robbers and tell them that if they were seen in Ogchoktu they would be shot at sight. The same evening Ristvedt prepared a little mine at the door of the provision tent, so arranged that it would burst when the door was taken away. There was no danger from it, but it would certainly have been enough to prevent any repetition of an attempt to break into the provision tent.

We fetched back both the boats which we had left in the fields. They had not suffered in any way.

Owing to an error in the laying down of my field magnet I found it necessary to repeat the work of the previous year around the station. Now in order to avoid losing a lot of time building snow huts, I let Hansen and an Eskimo go before me to the various places so as to build the huts for me. It was a great pleasure to find a house ready for me on arrival, so that I could commence my observations immediately. This observation trip occupied a month. The lowest temperature I observed during this winter at Kaa-aak-ka was nearly 50° below zero Fahr., consequently this winter was considerably milder than the last.

During the whole time we kept in constant communication with the Eskimo who had left us in February for seal fishing. Now and again some of them came to us and from time to time we visited them. We had a good example of the respect with which we had inspired these people. As may be remembered, the Ichyuachtorvik Eskimo had stolen the sledge and provisions from Rist-

Chapter VII.

vedt and myself when we were on our way to the magnetic pole the previous winter and established a depôt to the north on the coast of Boothia Felix. This year they came and gave up the sledge, being afraid that otherwise we might inflict some harm on them. They kept their distance, however, and put the sledge out on the ice near the camp of the Nechilli Eskimo. Their fear must have been great when they gave up such a valuable object as a sledge. It was, however, considerably knocked about, but under Lund's careful treatment it was made stronger and better than ever. It often happened that the Eskimo visited us late of an evening, and as I was sorry to see them go off in the very cold weather I invited them to pass the night lying in the cabin with the Lieutenant and myself; sometimes we had as many as thirteen guests for the night. They lay on the bare floor with deer-skins over them, placed together like sardines in a box. The occupants of the fore-cabin refused to have any night guests. They turned up their noses and declared that the smell was too strong for them. Neither I nor the Lieutenant ever thought they suggested either Eau de Cologne, violets, or new-mown hay when they took off their kamiks (foot gear) at night, but we were putting up with a little inconvenience, in order to proffer hospitality, which was a simple matter to us and also inexpensive.

In the course of the winter Lieutenant Hansen, Wiik, Ristvedt, and myself founded a Society, the object of which was, as far as possible, to taste all the products

The Second Winter.

of the land. Ristvedt was made cook to the Society, as Lindström "would rather throw himself into the sea than prepare such stuff." A fox steak, which the Society treated themselves to one evening, nearly brought our dear cook to the verge of insanity. He said we were the greatest pigs in the world, but the Society unanimously passed a resolution declaring fox steak to be one of the finest dishes that had ever been served on board. The flesh of a white fox, of which there were a great number, was really very fine, and reminded us of hare. We also tried many other dishes, such as reindeer tripe, and the like.

Amongst our equipment, I brought with me from home a great number of games, in all about seventy-five; of these, however, only two were used—"Fortress Game" and draughts—and then not by any members of the Expedition, but by the "Owl" and Talurnakto. Lieutenant Hansen had given them a slight idea of the game, what the object of the game was; apart from that they had not the faintest idea, but they assumed a very intelligent look, and amused themselves with the pieces for hours together. We laughed very much at them on the sly. No one ever knew how the games ended, but the players were quite satisfied. Their favourite diversion when they visited us on the vessel was to look at the illustrated books. At first they generally set the pictures upside down, but with our assistance they soon got used to the proper way of looking at them. Now, as luck would have it, we had hardly anything else but

Chapter VII.

pictures of the Boer War, and of these we had a large supply. It was death and killing and fire and slaughter, not very pleasant even to us, and the Eskimo impression of "civilisation" derived from these pictures can hardly have been happy and alluring.

The winter was now over, and there was every sign that spring was about to set in. There was a great difference between now and last year, as then it was about −40° Fahr. at the end of March, whereas this year it was about 17° F. This promised well for the summer, which would be of such great importance for us.

All the winter's work was over, and the station again surrounded by a circle of observatories. The long-distance sledge men were ready to start as soon as the weather was fine enough. This expedition was provided with food for 75 days, assuming that the depôt which had been established the previous year on Cape Crozier was still in order, otherwise there would only be enough provisions for fifty days. The expedition started on April 2nd, 1905, with mutual good wishes. They set their course towards the west, to Victoria Land, accompanied by our best wishes for luck and success.* This year we calculated that spring started when the expedition set out, although we had heavy snowstorms for some time; nevertheless, the cold of the winter was broken and did not return.

---

* See Appendix.

CHAPTER VIII.

PART I

THE INHABITANTS AT THE MAGNETIC NORTH POLE.

IN this chapter I will try to record my personal impression as to the Eskimo we met off the north-east coast of America, my observations as to their mode of life and their struggle for existence. Our individual impressions as to these Eskimo varied so greatly that I may say the seven members of the "Gjöa" Expedition formed different opinions of them. We had daily disputes as to their language and pronunciation, in fact we could hardly ever agree about a single word. If any other member of the Expedition were writing about these people, his description would no doubt be at variance with mine in many points, and it would be hard to say which would be the accurate version.

In undertaking to speak of these inhabitants at the Magnetic North Pole, the Nechilli Eskimo, I will therefore try to picture them as I found them and knew them. There are many sources of information and authorities on this subject which I could have consulted in order to give my readers a more complete chapter on the Eskimo, but I have intentionally abstained from

Chapter VIII.

reading them, simply because I feared I might be recording what others, and not I myself, had seen and experienced among them. Owing to my defective knowledge of their language there may, naturally, be some mistakes in my comprehension of the statements made to me by the Eskimo, but I think I may safely say that it is correct in its essentials. Our alphabet lacks letters to reproduce some of the sounds used in the Eskimo tongue. I have, therefore, taken such letters as would approximately correspond and tried to indicate the sound as nearly as possible.

According to their own statements, the Eskimo of the Boothia Felix Peninsula on the north-east coast of America and west up to the Coppermine River are divided into the following tribes : the Ichyuachtorvik tribe, who have their principal seat around Elizabeth, Victoria, Felix, and Sheriff Harbours on the east coast of Boothia Felix—the Ichyuachtorvik district, where Sir John Ross wintered in the "Victory," 1829—1833 ; the Nechilli tribe on Willersted Lake (Nechilli) on the Boothia Isthmus ; the Utkohikchyallik tribe in Utkohikchyalli or the regions around the outlet of the Back or Great Fish River into the Polar Sea ; the Ogluli tribe in Ogluli, the west coast of Adelaide Peninsula, and the Kilnermium tribe in Kilnermium or the regions around the mouth of the Coppermine River.

Through social intercourse, inter-marriage, and adoption these different tribes have mixed with each other to such an extent that they may now be practically looked upon as forming one single tribe. Their appearance, dress, customs, and habits are almost identical, and still further serve to promote their amalgamation. But I believe it is wrong (as is done in some reports and books) to regard all these different tribes as belonging to one original parent tribe, viz., the Nechilli.

Of all these tribes the Ogluli Eskimo was the one that had been most in contact with white men. In their districts many members of the Franklin Expedition breathed their last. It was also this tribe, that M'Clintock, Hall, and Schwatka met, while on their search for Franklin documents, and they were also the first we encountered. Several of them could remember the members of the Schwatka Expedition, and they still retained among them some remnants of English words such, for instance, as "oata" (water) and "naiming" (knife). Another striking proof of the fact that this tribe had been in contact with "civilised" men was, that old Teraiu, the very first night we met him, readily offered us his wife, Kayaggolo. Otherwise there were not among them any visible tokens, such as are usually found, of their contact with white men. Thus we found exceedingly little iron in their possession. Among the other tribes, except the Nechilli Eskimo, there was a complete lack of iron articles. Of the other tribes, the Ichyuachtorvik and the Nechilli Eskimo had been in contact with Englishmen from 1829 to 1834, but all who were then living had since died , there were only three old women of the Nechilli tribe who could say they had seen white men at Eivilli (Repulse Bay), where they had been with their husbands.

In the same year in which we arrived in King William Land, four Nechilli Eskimo had travelled south to Eivilli with skins to barter. This was a sign of enterprise we did not observe in any of the other tribes. Otherwise we did not see anything on our first meeting with the Nechilli Eskimo which would suggest any intercourse with the outer world, with the exception of some few iron bars and knives they had obtained by barter from Eskimo tribes dwelling further south.

We were suddenly brought face to face here with a people from the Stone Age : we were abruptly carried back several thousand years in the advance of human progress, to people who as yet knew no other method of procuring fire than by rubbing two pieces of wood together, and who with great difficulty managed to get their food just lukewarm, over the seal-oil flame, on a stone slab, while we cooked our food in a moment with our modern cooking apparatus. We came here, with our most ingenious and most recent inventions in the way of firearms, to people who still used lances, bows and arrows of reindeer horn. Their fishing implements were long spears, fashioned out of reindeer horn ; hour after hour they had to stand, heedless of fatigue, and waiting their chance to spear each single fish as it came along, while we put out our net and caught as many as we liked. However we should be wrong, if from the weapons, implements, and domestic appliances of these people, we were to argue

NALUNGIA AND ATIKLEURA OUTSIDE THEIR TENT.

that they were of low intelligence. Their implements apparently so very primitive, proved to be as well adapted to their existing requirements and conditions as experience and the skilful tests of many centuries could have made them.

Nechilli, the Nechilli Eskimo home and paradise on earth, is situated on the Boothia Isthmus, as already stated. The large Willersted Lake, with its moss-covered banks and its little fragment of a river flowing out into the sea, has borne the name of Nechilli for centuries. Here is the native land of the Nechilli tribe; on these very banks their fathers and forefathers have hunted from their tents in the light summer nights. Here they ran about while children with their little bows and arrows and shot at small birds, to be able later on to pursue and kill bigger game in hot chase. Here, in their youth, they had accompanied their parents and received much good advice and many a lesson, until in the end they themselves, as husbands and fathers, profiting by this advice and experience, took up in earnest the struggle for existence in the sphere of life allotted to them.

We are now in the middle of June, the most beautiful season in these regions. We may enjoy the splendid summer evenings in peace and comfort, as the gnats—this worst of all Arctic plagues, capable of making the loveliest summer night a very inferno—have not yet come. The tents are lying scattered singly all over the territory. One has selected a little knoll, with a good view, for the site of his tent; another the bank of a little lake, where the trout are big and fat.   The tents are no triumphs of art.   Most of them are made of reindeer and seal skins sewn together.   The best seal catchers have theirs made entirely of seal skin, and the worst have tents entirely of reindeer skin.   Seal skin is more precious than reindeer skin.

Atikleura has chosen the highest knoll for the site of his tent this summer, so as to be able to keep a better look-out for the reindeer when they come.   His tent is a model one ; it consists of thin, transparent, well-cured seal skin.   Even in the seaming it is superior to the rest. It has been set up with the opening away from the wind. A wealthy man like Atikleura requires three tent poles, one main pole to support the whole tent, and two placed cross-wise to form the entrance.   There is no special patent for a tent-door.   To hold down the tent, a ring of large stones is placed on the edge of the tent cloth. When the Eskimo move away, these stones are left lying almost in a circle ; these are called "tent-circles."   We found these tent circles scattered all over King William Land.   In the selection of the stones the Eskimo display some taste.   Thus, for instance, it is a pleasure to see Atikleura's tent circle.   His furniture does not consist of articles of luxury, but, for an Eskimo's tent it is remarkably clean and tidy.   The moss-covered ground on which it stands is thickly covered with reindeer skins, and as Nalungia, his wife, takes a pride in keeping the tent smart, the reindeer skins are neatly arranged.   There is no trace of any cooking appliance inside.   The kitchen

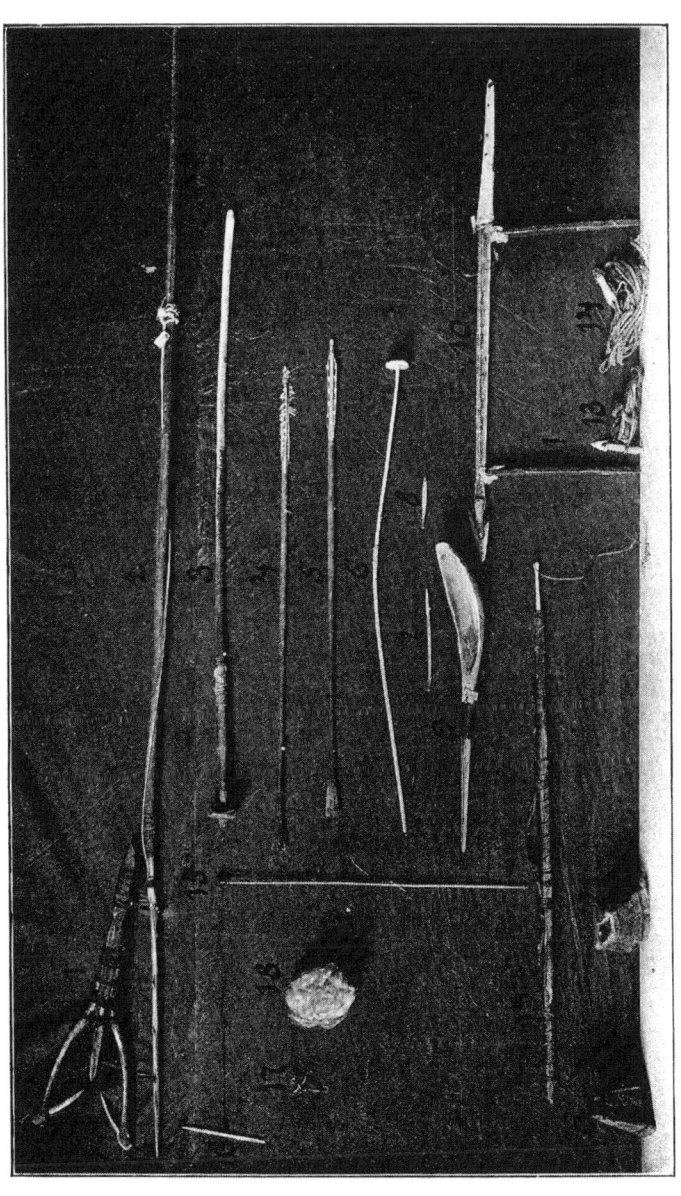

ESKIMO IMPLEMENTS.

is just outside the entrance to the tent. The blubber oil lamp is only used in winter; in summer they heat their food with heather. The fire is made between two stones, and the cooking vessel is placed upon the stones over the opening between them. The cooking vessels are of various sizes. In a model household like Atikleura's, it is about 2 feet 6 inches long by 10 inches wide by 10 inches high. It is made of a soft kind of stone, which they obtain from their friends the Utkohik-chyallik Eskimo.

Little Anni is playing around the tent. He is a spoilt child. Other children, of his age, are expected to work and make themselves useful, but it is something to be the son of a grandee. Errera, the eldest, is a fine lad. He is just returning home from fishing. He has his "kakiva" (fishing spear or trident) in his hand, and his catch is in a sealskin wallet on his back. He throws the wallet to his mother, who sits inside the tent busy with her everlasting sewing, and lays down his spear by the side of the tent. The fishing spear (Fig. 1) consists of four parts. The shaft is of wood, the longer the better; it ought not to be less than about ten feet long. The three pieces forming the trident are securely lashed to the lower end with cords made of reindeer sinews. Two of these parts are alike and form the outer prongs of the trident, each being barbed on the inside. The third piece consists of a sharp point driven into the end of the shaft and securely lashed. The fish is impaled on this point and prevented from slipping off again by the

Chapter VIII.

barbs on the two side pieces. Originally this implement was made of reindeer horn only; at the present day the central spike and the barbs are generally of iron, and the side pieces of musk-ox bone. The horn of the musk-ox is more flexible than reindeer horn.

Little Anni has approached his mother, who is turning over the contents of the fish bag to examine them. Now is his chance to sneak a tit-bit! The eyes of the raw fish are the object of his desire. These are looked upon as a great delicacy, and little Anni is not content till he has had his share. The other members of the family consume a portion of the fish at once; another portion is put in the pot to be "warmed." Nothing is cooked. Nevertheless this "warming" is not quite such a simple process as it sounds. They know nothing of our matches which light a fire in a twinkling, and their "tinder-box" is of the most primitive kind. It consists of two pieces of wood, one flat with a series of holes in it, the other in the shape of a pin, rounded off at either end, with a piece of reindeer bone, and a thick strong cord of reindeer sinews. In addition to these a little bag of moss, dry as tinder, is indispensable. To light a fire the flat piece of wood is placed on a hard base, with the holes upwards. Then one end of the round pin is inserted in one of the holes, the other end being firmly held in a corresponding hole in the piece of reindeer bone which rests against the operator's chest. Now the reindeer-sinew cord is wound once round the pin, and the operator draws the cord rapidly backwards and forwards as though drilling a hole. This is slow work in the winter, but in dry weather, like the present, it does not take more than a few moments to produce fire. Smoke is soon seen rising from the two pieces of wood, and the operator stops to inspect progress. If he has "drilled" long enough he has produced a fine wood dust which is now smouldering. The smouldering dust is knocked from the piece of wood into the bag of dry moss, where the fire is fanned by vigorous blowing. When the moss is well alight other fuel saturated with train oil is added, and this flares up immediately. The one who is first to get a fire is always willing to oblige his neighbour with a "light."

On one occasion I showed them how to get fire with the sun's rays through a lens. They were much amused, but no idea of possessing a lens and using it ever occurred to them. The fish is now warm, and the whole family pounces on it. Of course, knives and forks are unknown to Atikleura and his family. They content themselves with the eating implements Nature has provided. The food may or may not be warm, they are not over-parti-cular. After the meal the family busy themselves with a little work of various kinds. Errera finds that his fishing spear is not quite as it ought to be, and sets to work to adjust it. Eskimo always sing while at work, if I may describe the sounds they emit as "singing." But not even the most trivial duty is ever performed by man or woman without the accompaniment of these strange, monotonous vocal productions : c—d—e—f, f—e—d—c, c—e—d—f, d—f—e—c, and so on, *ad infinitum.* Even to one whose highest musical achievement consists in singing a

KAYAK IN COURSE OF CONSTRUCTION.

nursery rhyme out of tune, this monotonous music was maddening. However, when I visited them they always stopped singing, as they knew their unmusical performance at once incited me to imitation, and rather than hear me, they lapsed into silence.

While Errera is repairing his fishing spear, Nalungia is making preparations for getting Atikleura's "kayak" ready. This is now lying outside the tent on two small piles of stone, sufficiently high to prevent the dogs from getting at it. The woodwork is nearly finished, and all that is required is some final lashing and glueing. It has been a work of patience to construct the kayak, as it consists throughout of quite small pieces of wood lashed together. Reindeer sinews are used for lashing; they are excellent while dry, but when they get wet they stretch and the lashing slackens. It is only where a superior mode of joining is required that glue is applied. This is made from reindeer blood by a very peculiar process. Having filled a small bag with blood, the Eskimo takes it in his mouth and gently sucks at it for a long time. By this treatment the blood coagulates and forms a thick liquid, which is equal to our best glue.

Atikleura is now anxious to have his boat finished and in full working trim by the time the ice breaks up on the river, as the reindeer have already begun to arrive in herds, and the great summer hunt will soon begin. A little fresh reindeer flesh will be very acceptable; it is so long since they had the last. Besides, the housewife's stock of thread is nearly exhausted; there will be none left when she has finished covering the kayak; and seal sinews are no good. It is quite a pleasure to watch her, as she is now hurrying to finish. The single thread she has is too thin for fastening the kayak cover on, so she is busy plaiting a stouter one. Her little hands—indeed it is rare to see finer and more shapely hands and feet than those of the Eskimo women work so fast that the eye cannot follow the movements of her fingers.

In this way the family is fully occupied. We will now call on Atikleura's neighbour, Kaa-aak-kea, to see how things look there. It is difficult to understand why he chose this particular site for his tent, as it is neither on an elevation nor is it near water. But eccentricities are to be found everywhere in the world, why not among the Eskimo? You can see with half a glance at the tent that it is the complete antithesis of Atikleura's. Probably there is not a single sealskin in the entire tent cloth. It is made of reindeer skins, full of holes and worm-eaten, sewn together. He has not even troubled to remove the hair properly; large tufts of it still show here and there, and the sewing is on a par with the skins, just a stitch now and then at long intervals. It is so badly done that I believe, even I, could have managed it better. A quantity of fish bones are heaped outside the tent attracting swarms of immense bluebottles. "Manik-tu-mi, Kaa-aak-kea!" I call out. A long-drawn "manik-tu-mi" answers lazily from the heap of skins inside. He is not up yet. All that is to be seen of him is a tuft of towzled hair and a pair of blood-shot eyes, protruding from a skin cover. All within is filthy and malodorous, so I do not stay very long. I can hardly sympathise with Kaa-aak-kea as it is due to his own brutality that he is now living lonely and forsaken. He frequently ill-treated his first wife, beat and thrashed her till she was black and blue. She died in her last confinement; a happy release! However, Kaa-aak-kea found bachelorhood rather dreary and soon looked round for another wife. But this is no easy matter in a world where womenfolk are so sparsely represented as among these Eskimo tribes. He, therefore, conceived the sensible idea of going further a-field, to one of the other tribes, to seek a wife. Some months later he returned, bringing back with him a child of nine years. This child was to be his wife! What their relations really were I could never satisfactorily make out. He, himself, said that she was not to be his wife for some years, but I had my own opinion on the subject. It was shocking to behold how this child was clad. She went about in some old worn-out clothes of Kaa-aak-kea's which, of course, were many sizes too large for her. Her foot gear was much the same. She got many cuffs and little food. But one fine day she decamped. Hungy and almost destitute of clothing, this child walked thirty miles, until she found people who took pity on her. Now the ogre is living alone, with his two little boys.

The elder brother Akla, was many degrees better than Kaa-aak-kea, though even he could not be said to be an ornament to the tribe. He was married to " Pandora," the biggest and strongest lady of the tribe. Probably it was on account of her possessing these properties that Akla refrained from indulging in the ill-treatment Kaa-aak-kea

Chapter VIII.

inflicted on his wife. This marriage was a typical "happy marriage"; she reigned absolute and he obeyed blindly. It was this lady who played such a momentous part in the life of our friend Talurnakto. It was her charms that enticed him, for the first time, from the path of duty, and the substantial nature of these charms left its traces on the poor wretch during the time he played

OYARA AND ALO-ALO BY THEIR SNOW HUT.

the part of second husband in Akla's house. "Les extrémes se touchent." Talurnakto was the smallest man of the Nechilli tribe ; Pandora was the Amazon. As a matter of fact Pandora's was not a model household. The youngsters were filthy and repulsive, and Akla's clothing was not always beyond reproach.

The third brother, Oyara, was the best of the lot, and one of the most attractive types in the tribe. He was about twenty-seven, tall, dark, and with a friendly winning smile. He and his bosom friend, Ahiva, were among the swells of the tribe. They both had young, sweet wives, Alo-Alo and Alerpa. When I first met these two couples, Oyara was married to Alo-Alo, and I do not remember having ever seen a more loving couple. When I met them again, however, a little later, the conditions had changed. Alo-Alo was now occupying the housewife's place in Ahiva's tent. When I intimated that I must either have been, or was now under, a misapprehension, the little coquette lent towards her husband, took his head between her hands, and rubbed noses with him, as much as to say : " Look here, there can be no mistake about this." " Rubbing noses " among the Eskimo is equivalent to " kissing " among Europeans. Of course, I took in the situation at once. This sort of thing, in fact, is not quite unknown in " Kablunaland " either, but it was the only case of exchange of wives I had actually witnessed. Both of these men were smart huntsmen, and their wives, young as they were, were capable housewives. Their tents were situated near each other, in a charming little spot, where, I should say, love could thrive.

On a prominent point down near the river lies another tent. It must be owned by some " great man," as it is large and elegant. He, too, has done well in fishing, and judging by the long rows of fine small salmon and trout hung up for drying, it seems he has an eye to the future, which is not generally the case with

Chapter VIII.

the Eskimo. The laughter of the owner, Kachkochnelli, can be heard a long way off. He has reached
mature age, between forty and fifty, and is already a
grandfather. He cannot be called handsome, rather
the reverse. He is of medium height, rather stout, and
red-rimmed eyes are his most distinctive feature. He is
always full of fun and high spirits. Kachkochnelli is an
intelligent man, and, in fact, one of the most capable
huntsmen, as well as one of the sharpest business men
of the tribe. He has always some business transaction
on hand. He has a wife and three children. His wife
looks like an ordinary, well-nourished peasant woman ;
she is always quiet and friendly. She passed among us
under the name of " Nuyakke," or mother-in-law. This
was because Kachkocknelli wanted me to buy Oyara's
wife Alerpa, who was his daughter ; his wife would then
have been my mother-in-law. This proposal, of course,
caused general ecstacy all along the line. But the
transaction was not completed. He also was always
on the look out for a purchaser for his wife, but in this,
too, he was equally unsuccessful. His eldest boy, Kallo,
was about ten, a pretty but mischievous lad. His
youngest son was named Nulieiu ; he was one of the
queerest youngsters I have ever seen, always disgustingly
filthy. But it was his clothing that attracted most
attention. Like all little Eskimo children of his age,
about six, he was dressed in a suit made in one piece,
that is to say, trousers and jacket in one, a kind of
" combinations." Nulieiu's dress, however, was dis-

tinguished by being particularly airy, inasmuch as it was open in front from the neck down to the stomach ; while the other details of its construction were such that it was constantly pervaded by a thorough draught. I saw him in this airy dress until the severest winter cold set in. In spring they were more careful with him, but when the temperature went up again towards freezing-point, he would reappear in this summer " reform " dress. When Eskimo children are over six, the dress is closed in over the whole body.

In the small tent a little further off dwells Poieta, Atikleura's younger brother, with his wife Nalungia and their little son. They have not much space inside, but it is clean and neat under the circumstances. Nalungia is devoting her attention to her little boy ; it is very touching to see her treating him exactly as a cat does her kitten, licking him all over the body, and there he lies as clean as a new pin. The boy, in fact, is not her own ; she received him as a present, from a mother who possibly had too many. After this bath the child is thirsty, and proclaims the fact by squalling aloud. She at once puts him to her breast. The little one, thinking his thirst is going to be quenched, becomes quiet at once. But Nalungia has never been a mother. And now she performs one of the most amusing tricks I have ever seen. She knows, of course, that the boy will not submit long to being tricked, so she thinks it best to be beforehand with him. She takes a mouthful of water, and, before the child has time to commence shrieking,

Chapter VIII.

she brings its mouth up to her own with a movement quick as lightning, and with astonishing skill performs the trick of letting her boy drink the water from her own mouth. Seeing visitors approaching she puts the child down, so as to have a little chat with the newcomers. The child is meanwhile lying snug and contented under a reindeer skin and is sucking his "comforter." The "comforter" is neither a bottle nor a sugar bag, but is simply a piece of blubber through which a long skewer is drawn. There is no fear of the child getting the blubber into his throat, as the skewer will bar the way, and the child may therefore be safely left to himself. Nalungia is a stout, pleasant-looking woman of some two and twenty years, red and white like a rose, and looking the picture of health, yet withal somewhat sluggish in appearance.

It is no enviable position to be a mother in these regions. Up to the time the child is two years old, frequently even longer, it is carried on the mother's back in a small slit made in her clothing. It is not carried as one would imagine, in her hood. This hood is merely an adjunct to the female garb, but has no special object, except that it is put over the head in cold weather ; it would be equally serviceable for this purpose if it were only one-third of its actual size, but designed as it is, it might be a very handy place of concealment for stolen goods. The child is always carried in this kind of slit or pocket-hole, which is so small as to be scarcely noticeable when unoccupied. The child lies in it with his legs drawn up, like a little frog, quite naked, warmed by the heat of the mother's body. To prevent the child from slipping out, the mother has a cord of reindeer sinew slung round her over her outer clothing. The cord is fastened in front on the breast, with the aid of two wood or bone buttons, and can be unfastened in a moment to remove the child. This has to be done pretty frequently and sometimes with alacrity. The poor unsuspecting infant is brought out into the open air with marvellous speed. I have seen quite little babes on these occasions brought out from their warm resting place and kept for several minutes, perfectly bare, in a temperature $58°$ below zero (Fahr.). One would think no infant could stand it, but to all appearances it does not seem to matter much to the little Eskimo.

There is a little romance attached to the marriage of Poieta and Nalungia. Like most of her Eskimo sisters, Nalungia had been allotted to her intended husband from her birth—not to Poieta by the way. It so happened that Poieta fell violently in love with her after she was married. He made short work of it. He simply went to her abode and carried off the bride, presumably with her own approval and consent. When the husband came home he found the house deserted, and his demand to get his wife back was answered with a contemptuous "Come and take her if you dare!" As the husband lacked courage to do so, Poieta remained in peaceful possession of Nalungia. The deserted husband migrated southwards, and report has it that he fell through the ice and was drowned.

Chapter VIII.

Weddings are celebrated quietly and without much ado. When the girl is fourteen she goes to her future husband, or, perhaps, he goes to her, and she lives with him in his parent's home. I do not think that, as a rule, any particularly romantic feelings animate the Eskimo in this matter, either on one side or the other. The

Kabloka.   The "Owl."

A LIVING BOOTJACK.

women marry simply because they are given away by the parents, and the man marries the woman to procure another domestic animal. Such is the wife's position, neither more nor less. Even our good friend, the "Owl," who has his tent on the other side of the little valley, right to the west, is not much better in this

The Inhabitants at the Magnetic North Pole.

respect than the rest. He is just returning home with his bow and arrows—he has been bird shooting. Without a word he throws down the birds in front of his little wife, Kabloka, who is sitting there at her sewing. See sees at once what is required of her, and without demur begins to open the birds. Meanwhile the " Owl " sits down to rest, after his long day's tramp. The first thing is to remove from his feet his wet, muddy sealskin boots. He would be no Eskimo if he did it himself. He simply stretches his legs towards his wife, and she at once lays her work aside and lifts up his foot with both her hands, gets her head under it, secures a good hold of the back of the "kanik " (shoe without heel) with her teeth, and pulls it off. She does not mind getting her mouth full of the mire and filth through which he has trudged in the course of the day.

I can now understand why the Eskimo rub noses instead of kissing. The mouth, besides being a very good talking apparatus, serves a variety of purposes ; it is the Eskimo's universal tool. It is wonderfully well developed, large and powerful. Their teeth are of peculiar shape. Ours are pointed and thin ; theirs have a large broad masticating surface. They wear their teeth quite down to the fangs, which, of course, is quite unknown with us. I never heard toothache mentioned among the Eskimo. Anna, the " Owl's " mother, apparently still retained her teeth intact, but they were worn away almost down to the gums. If there is anything that the fingers are not strong enough to perform,

Chapter VIII.

the teeth are brought into requisition ; thus, to straighten a nail is a mere trifle to their teeth ; they cannot manage it with their fingers.   In the " Owl's " tent, Anna rules absolute.   The old lady is, in fact, a kind mother-in-law, and Kabloka is, therefore, fond of her.   She can still do all the sewing required for the family.   Kabloka is too young, and has yet much to learn.

The Eskimo have no marked sense of order.   When they have done sewing they put down their needle anywhere.   When they want it again, some hours after, there is a hunt for it.   It is like looking for a needle in a haystack.   All the skins must be moved, and everything turned upside down before the needle can be discovered.   They were very short of needles before we arrived.   To be possessed of a needle is considered positive wealth.   It must be remembered that a needle represents a large amount of work, for the paterfamilias has manufactured it out of a piece of iron or copper, whichever he could get hold of, and the workmanship amply proves his skill, as with the most primitive tools, to produce not only a serviceable, but a really good needle, is certainly no easy task.   It is true the Eskimo has plenty of time for it, but I should like to see anyone among us who, even with unlimited time, could accomplish a task requiring such skill and patience.   The Eskimo is endowed with a large share of manual skill and practical sense.

Alongside the " Owl's ' tent stands the abode of his brother, Umiktuallu, and his wife, daughter, and little son. Umiktuallu is a great seal hunter, and is quite on a par with Atikleura; therefore they, too, do not want for anything, and have plenty of rugs and clothing. There is, it need hardly be said, a villainous odour of fish in all the tents. All one sees and touches is impregnated with fish-slime. The boys fish for cod on the ice, and the adults fish for trout in the lakes. Cod is caught with the line, the hook consisting of an old bent nail. A little urchin of five or six will bring home a large catch of cod after a day's fishing. They have no regular night-sleep at this time of the year ; they lie down for a nap whenever they feel inclined at any part of the twenty-four hours. It is as light at night as it is during the day. There is trout fishing in the lakes, in the open water along the banks. For catching trout they use the " kakiva," or fishing spear, already described, and " lokker " (ground bait). A little later on, when the ice breaks up in the rivers, they fish in the small torrents for salmon, and kill large quantities with the " kakiva " alone.

There are still four tents belonging to our acquaintances which we have not yet visited. In one of these lives Keyo, with his wife, Nalungia, and a little daughter of eight called Kamokka. Keyo would certainly have been an excellent fellow had it not been for the temptations of civilisation. I regret to say it, but after coming into contact with us he was no longer what he ought to have been. Whether the fault lay with him or with us I will not venture to say ; probably there was blame on both sides. From being truthful and industrious, he had degenerated into a mendacious, lazy rascal. He simply lounged about doing nothing. His wife was a very nice woman, considerably over medium height, slender, and more like a " Kabluna " than any of the others. It was sad she should have such a worthless wretch for a husband, for, as far as I could learn, she was very well disposed, but his treatment of her was that of the vilest bully ; he commanded and she obeyed.

Two of the other tents were occupied by Kirnir and Amgudyu with their respective wives. Kirnir was an Ichyuachtorvik Eskimo, but had camped with the Nechilli for the summer. He was a man of twenty-five. His wife, Magito, was about twenty, and was very handsome. Kirnir was an intimate associate of Keyo, and treated his wife much in the same manner as the latter did, but as little Magito did not seem to take it amiss I could scarcely pity her. The third couple were Angudyu and his wife. Angudyu was tarred with the same brush as the others, a scamp of the first order. His wife, Kimaller, was about twenty-two, and, in my opinion, exceptionally good looking. That such a woman should have fallen into the hands of such a scamp was a very great pity. There was something attractive and smart about Kimaller. She was always very quiet, rarely laughed, but had an exceedingly winning smile. Her handsome eyes with their deep sad look made her very engaging. She really possessed, what I very seldom found in other good-looking repre-

The Inhabitants at the Magnetic North Pole.

sentatives of the fair sex among the Nechilli Eskimo—
grace. These three men were known among us as the
" bullies."

Lastly, in the fourth of these tents, lived Nulieiu, the
Ogluli Eskimo, and he was the very antithesis of these
last three personages. He also had settled here for the
summer with the Nechilli. He was between thirty-
five and forty, short set, and with a rather fuller growth
of beard than most of his comrades. His promise was
to be absolutely relied on. I cannot help giving a little
instance which is quite typical. He was among the
Eskimo as our guest in November, 1903. When he
left, Lieutenant Hansen commissioned him to get several
articles of clothing for him. In the autumn of the
following year he returned and brought with him every
single article ordered. During the time we were
staying in Ogchoktu he brought us large quantities of
reindeer and fish. He was capable and honourable in
every way. I admired him particularly for his treatment of his wife, and in this respect he stood out with
distinction among the other Eskimo. In fact, I believe
that he was the only one among them who did respect
his wife. She was about thirty, and by no means
unattractive. Her name was Kayaggolo. They had
a little daughter about eight years old. Nulieiu was
a declared friend of the Expedition,

Talurnakto, who is always on the move, living now in
one place, now in another, has been so fully referred to
before, that there is no need to describe him here at

Chapter VIII.

length. But there is one person I have forgotten ; he is just coming up the slope now ; he has been absent during this lengthy round of introductions. He is Aleingan, the oldest member of the tribe, and their champion charlatan. He is also known as " Kagoptinner," or " the Grey-haired One." He reminds me of a sly old fox. He is the father of Atikleura and Poieta, and lives with the latter. Between sixty and seventy—his black hair and splendid beard, now much mingled with grey, hence the sobriquet " the Grey-haired One "—gruff and imperious in manner, I really believe he commands great respect among his people. He is very strongly built, and doubtless was once a powerful man. He has the reputation of being an " Angatkukki Angi " or magician, and in token of this, always wears a large collar with fringes of reindeer skin over his shoulders. He regards himself as the chief of the tribe, and his demeanour is shaped accordingly. I was never quite able to make out whether he really considered himself superior to the others, or whether it was all mere acting. I am strongly inclined to believe the latter. He had pursuaded the whole tribe that he had been on the moon, and gave them long descriptions of that planet. Of course, he himself had occupied an important position up there. Those Eskimo who conducted themselves on earth in an exemplary manner, that is to say, who did as he told them, would, so he assured them, find a place up there. Reindeer hunting was carried on up there on a large scale, and there were many other forms of enjoyment. Those who had only been partially obedient to him were assigned a place in the stars, where, of course, he had also been, and those who had always refused to obey him, would be cast down into the earth. This they all fully and steadfastly believed. When I once treated the " Owl " for a broken collar-bone, they would not be satisfied until Aleingan had had a hand in the matter. What this consisted of I do not know. Aleingan was one of the few Eskimo who had two wives.

There was in this tribe an ancient tradition as to a race of giants who had once lived in this land before they themselves came here. These were called " Tungi," and were always spoken of with the greatest respect. They were said to have been considerably taller than the Eskimo and much stronger. Bearskin was their clothing. Some ruins of ancient stone huts which we found in the neighbourhood of Ogchoktu were supposed by the Eskimo to have been Tungi huts. Old Aleingan, among all his other rigmaroles, would also relate that he had slain the last of the Tungi. And all believed him ; no one was inquisitive enough to ask him to let him see the remains. For the rest I saw nothing of his skill ; he was prudent enough to keep his talent concealed from a discriminating public. I could never persuade myself to consider him other than a consummate old humbug who traded on the gullibility of his fellow-tribesmen, and I, therefore, counteracted his influence as much as I could, though I think it took me a long time to make any impression at all.

Chapter VIII.

Meanwhile, time is advancing ; spring is approaching with huge strides, and the season is at hand when the kayaks are to be covered. They have been standing several days, quite ready as far as the men's work is concerned, though the long preparations required in dressing skins and making thread have delayed the work of the women, but they are ready now. Covering a kayak is no easy matter ; it is very laborious work and one woman cannot manage it alone. She has to get help from her female friends, and they render it willingly, knowing they will need help themselves in their turn, as they all have to cover kayaks. In covering a kayak, only the skins of full-grown female seals are used. Six skins are sufficient for one kayak. In dressing them, the skin is stripped off, and scraped as clean of blubber as possible, with the " olo " or woman's knife. This operation is best done while the skin is stretched out on a piece of wood. After that, the skin is chewed and sucked, to remove any small particles of fat which the " olo " has not taken away. After it has thus been carefully freed from blubber, it is rolled up with the hairy side outwards. The bundle is then rolled up again in a piece of skin, also with the hairy side outward, and the whole is laid down so as to be exposed to a gentle warmth from below. After a time the hair begins to fall, so that it can be scraped off with the thumb nail. When the skin is quite free of hair, each one is rolled up separately, and they are then all packed together in an ordinary sealskin and buried in the snow to freeze. When the summer comes and the snow disappears from the fields, the skins are taken up and sewn together. Then they are stretched wet over the kayak and sewn on. It is this stretching that requires strength, and it is then that the women need help from their friends. When the skin is sewn on, the kayak is left to dry. While drying, the skin shrinks and becomes taut like a drumhead, turning at the same time light yellow in colour and almost transparent.

As we are dealing with this important question, I may also refer to the other methods of skin-dressing in general. Unlike some Eskimo tribes, the Nechilli do not employ urine in their skin-dressing. For their " Kamileithun," or outer kamiks, they use the skin of one-year-old seals. The mode of dressing is the same as that employed for the kayak skins, up to and including the stage of chewing and sucking. But, instead of putting them into the snow to freeze, they are spread out to dry gradually in a gentle heat. When dry they are worked with the teeth until they acquire a suitable degree of suppleness. A skin dressed in this way is whitish-yellow, and opaque, in contrast to the translucent kayak skin. For the " epirahir," or fishing-boots, they use the skins of both male and female seals. They are completely freed from blubber with the " olo," and exposed to heat until quite thawed. When all the fat is removed, the woman stretches the sealskin over her thigh, and by means of the " olo," ground up to a razor edge for the purpose, she shaves off the hair cleanly down to the root. The

Chapter VIII.

practical Eskimo women must have found that the thigh is a soft and level groundwork for the skin to rest on, while at the same time it is yielding to the knife-edge. But it must be rather a chilly occupation when the skins are only just thawed. An industrious woman can shave three skins a day. The skin is then spread out in the snow and dried. But before it can be sewn it must undergo a thorough masticating process in order to become sufficiently supple and pliable. In spring, the Eskimo women may be seen munching and chewing these skins all day long. The skin of male seals, being the thickest, is used for kamik soles. This does not undergo any elaborate treatment; it is only freed from blubber, stretched, and dried. The skins of one-year-old seals are used for sealskin garments. Seals of either sex will do equally well. After the skin has been freed from fat, it is cleaned out in the snow. Tepid water is poured on the hairy side, and a quantity of snow is put on it; the woman now treads and tramples on the skin in this mixture of water and snow, until the hair is absolutely cleansed of all blubber. Then it is stretched out on the snow and dried. Later, when the skins are to be made up into garments, they are not shaved off, but the husband works the skin with his fists until it is quite soft. The skin of very young seals, under a year old, is mainly used for spring trousers. For tent skins both male and female seal-skins are used. Lazy people leave the hair on, but those who like to have everything neat and smart, scrape it off. When the reader learns that the seals caught by the Nechilli Eskimo are quite a small species of seal—snadd—he will understand that a large number of these skins are required to supply their wants. If we see a tent made of sealskin we may be sure that the owner is a great seal-hunter.

Reindeer skin is used for clothing, bedding, and tents. If for clothing it may be treated in two different ways. The first is simply by drying in the open air. After the reindeer has been skinned, the skin is spread out in the field, without stretching it, and secured by small stones laid on its edge, or by reindeer ribs stuck through holes made on the outer edge of the skin. When the sun shines, the skin dries rapidly. If we look at the hairy side of one of these air-dried skins, we cannot see down to the roots of the hair. Skins prepared in this way are only used for outer clothing for the winter or for bed rugs. The advantage of these skins is that the snow cannot penetrate through the hair, and consequently it is easier to keep them dry. Yet they are rather stiff to the touch even when they have been scraped with iron, stone, or bone scrapers, and made as white and fine as possible.

The second method produces a suppler and finer skin, chiefly used for underclothing, but also frequently for outer garments. After the skin has undergone a slight preliminary drying in the open air, preferably after snow has fallen, it is taken into the hut, and the Eskimo covers himself with it at night, placing the flesh side next to his own body. Over this he has the usual bed rug, with the hairy side turned inwards. One night is, in most cases, sufficient. In the morning the skin is rolled up and exposed to the cold so that it freezes stiff. Later in the day the wife goes over the flesh side of the skin with the shoulder blade of a reindeer to soften it, and at night, when the paterfamilias comes home from his day's work, he gives it the final treatment according to all the rules of art, with the usual scrapers. The freezing and softening process, after the one night's tender nursing, has an absolutely surprising effect on the skin. It is as soft as silk and you can see right down to the roots of the hair. It is not so useful for outer winter clothing as the other, because the snow gets in between the hairs down to the skin itself, where it melts from the heat of the body and makes the clothes wet. The skins of the reindeer caught in the spring are used for tents, because they shed the hair and are not serviceable for clothing.

Now the great season is approaching when the ice on the rivers and lakes breaks up and reindeer hunting begins. This is without comparison the Eskimo's most glorious time. He then gorges himself with meat. Reindeer begin to come north of the Boothia Isthmus as early as the month of May, but it is only now and then that one is killed with the bow and arrow at that season. It is only when the ice disappears and the kayak is launched that reindeer hunting begins in earnest.

When the kayak is fitted out for the reindeer chase, the hunter has his two reindeer lances ready beside him, fixed in straps in the kayak skin so that they may not fall overboard. The reindeer have a fixed track to the north for their spring migration. When passing Nechilli they take such a course as to make it easy for the Eskimo to drive them into the water. The huntsmen divide into two parties, one with the kayaks and the other without. The kayak men station themselves on the bank opposite to the one from which the deer are coming. When a herd of reindeer approaches, the drivers make a wide circle round them and drive them into the water. As soon as the deer are well into it, the kayak men jump into their boats and spear one reindeer after another. The animals are towed to the bank and taken care of by the huntsmen on shore. Later in the summer, when the reindeer have spread over the whole country, the hunt is as a rule carried on at one or other of the large lakes, preferably where a point of land projects. Then there are fewer Eskimo together and it often happens that when about to drive a herd of reindeer on to the point, they find they are too few in number to manage it. But the Eskimo are men of resource. They hastily build up a lot of small cairns, and when the reindeer come they mistake the cairns for men and the ruse succeeds. A small number of Eskimo will often manage in this way to drive large herds into the water, and when once they are afloat there they are doomed ; very few are lucky enough to escape death.

According to the statement of the " Owl," the booty is divided so that out of five reindeer killed four go to the

Chapter VIII.

kayak men and one to the huntsmen on shore. However, this does not signify much as the socialistic principles of the community do not permit any great accumulation in one hand. They are in the habit of feeding on the meat all together as long as there is any left. The skins and certain parts of the carcase, however, go inevitably to the one who killed the animal.

After a wholesale slaughter of reindeer there is first of all a great feast on the spot while the animals are being skinned. The Eskimo are very careful over this work, as every skin is precious. However, most of the carcases are laid up in a depôt, built carefully of sharp stones, to keep out the foxes, although it sometimes happens that these rogues steal in, and it is incredible what havoc they can work in a short time. In summer, of course, the meat soon deteriorates; but no matter, apparently it seems to go down just as well. As to "blood pudding" I have already discussed that article sufficiently; it is manufactured on a large scale during the summer season. The main things taken home from the hunt as soon as the animals have been killed are the skins, the thigh and shin bones, the fillets, tongues, and sinews. The last named are most carefully treasured. The sinews of the back are the finest, and are used for sewing thread. The other sinews are made into coarser lashing thread or cord, corresponding to our string.

When a hunter returns home with this, his booty, the sinews are at once hung up to dry. The meat is consumed forthwith; the marrow, which the bones

contain in large quantities, serves as a kind of dessert.
They are careful not to use any iron implements for
breaking the bones, as it would be unlucky : "the hunting
season would be a poor one." The bones left over are
carefully buried lest the dogs should get them, the
popular belief being that if they did the hunter would
not find another reindeer that year. The same belief
prevails as regards fish bones during the fishing season.
According to this it is pretty clear there cannot be much
left for the dogs.

After a meal, rest and relaxation is the order of the
day in the camp. The dogs lie down on the moss and
stretch themselves, always seeming to enjoy themselves
in spite of their empty stomachs. Laughter and jest
sound merrily from the tents out into the beautiful night.
The sun is nearing his lowest position and is just on the
horizon, casting long, ruddy beams over the landscape.
The large Lake of Nechilli lies smooth as a mirror ; the
kayaks, drawn up on the banks and turned upside down
on two high piles of stone, so that the dogs cannot get
at them, are still dripping from the sanguinary hunt.
Gradually the tents are closed down, one after the other ;
the Nechilli fall asleep, and the profound peace of the
Polar sunlit night is undisturbed, save now and then by
the ever restless owl.

If the summer is glorious it is also brief, and the
autumn soon sets in. As far as weather conditions are
concerned, there is little to be said about autumn.
Summer is, one may say, rapidly succeeded by winter ;

the lakes freeze over, and the snow falls; but with the
Eskimo there is a short period which may be described

THE NECHILLI IN SUMMER TIME.

as their autumn, and as their most dismal season, just
before the ice is thick enough to be used as building

330

material. Superstition prevents them from lighting fires indoors. Their homes are, therefore, miserable in the temperature which then prevails, and they live in a raw cold, damp atmosphere, in which all, without exception, contract severe colds. The hunting season is now over, though now and then a solitary reindeer may still be seen. The many large piles of stones seen all over the country indicate that the hunters have been fortunate, as every pile shelters a food depôt, and often contains several carcasses. Every family is now well stocked with skins and thread, wherewith to make clothing, and can look forward with equanimity to the coming of the winter. Most of them continue to live in the tents, and postpone building huts until sufficient snow has fallen. Some of them, however, do not mind the trouble of building an ice-hut at this time of the year.

The " Owl " was one of those few who thought more of his comfort than of the work entailed. His good friend, Talurnakto, had promised to help him, for an ice-hut such as this is difficult to erect single-handed. They start in the early morning, armed with long knives. They are in high spirits ; their merry laughter rings out afar in the stillness. These happy folks require no fortifying by stimulants like scalding hot coffee or tea to keep up their animal spirits ; a mouthful of icy-cold water is probably all they have taken. Soon they are down on the Nechilli Lake hard at work cutting ice. It is not very thick yet, not more than six inches, so that it does not take long to cut through ; then they both lie down on their stomachs and drink from the holes they have cut. I have never seen an Eskimo cut a hole in the ice without drinking from it. This is not done so much from actual thirst or, as many suppose, on account of the insatiable desire of the Eskimo for everything he sees, but from a kind of instinct to avail himself of the opportunity. He knows from experience that it may be a long time before he can procure water again, and thinks it best to take the opportunity when it presents itself.

The slabs cut out of the ice are about five feet by three feet. The Eskimo do not detach them entirely by cutting, they only make a number of holes in the ice, along the intended outline of the slab, and detach it by a kick or pressure. Talurnakto's greatest delight is to detach the piece thus cut out by jumping on it. He is so reckless that I should never be surprised to see him fall through. When the slab is entirely detached from the ice all round, a little hole is made in each corner and sealskin straps are passed through these ; the slab is then easily hauled up. Nine of these slabs suffice the " Owl " for his hut, and it does not take long to convey them to the " building site." Using a pail made of kayak skin, they make a mixture of snow and water, which binds more quickly than any mortar used by our builders. The slabs are now set on edge in a circle so that the five-foot side is the full height of the structure. The snow mixture is thrown in between the joints of the slabs, and it binds them up into a compact ice wall five feet high. A skin roof is stretched over this. All that is now wanting is the entrance. As I sit here, I am just thinking I should make it by cutting it from the middle of one of the ice slabs ; that would seem to me the most feasible way. But the Eskimo is too practical to cut up the good compact wall, he makes the doorway where there is already a natural break, that is, at the joint between two slabs.

A house like this is very speedily erected, and has the additional advantage that it can be occupied at once. Old Anana and little Kabloka need no second bidding to exchange the dark uncomfortable tent, where they have been huddling and shivering with cold, for the new, light, spacious ice-hut. If it were not for that stupid moon, they might have lit a comfortable fire and made the hut warm and cosy. But while the moon is in this particular quarter, venturing to light a fire is out of the question. Such superstitions as these abound among the Eskimo, and it is useless trying to convince them of their absurdity. For instance, during the hunting season, no sewing or stitching must be done, beyond the most indispensable repairs and no one could induce them to touch a needle except for some case of absolute necessity. One morning, when there was a very sharp frost, I saw my friend Akla sitting down sharpening his knife, having the whole upper portion of his body perfectly bare. His coat lay at his side, while he was shivering and numb with the cold. I went up to him and asked him what he was doing sitting there in that condition, and whether he fancied it was summer. And when I rated him for his folly, and told him to put on his coat lest he should catch his death of cold, he smiled sadly, but left the coat lying where it was. Later I learnt that it is not considered proper to work with iron, for some time after the death of relatives, without baring the upper part of the body, and he had lately lost both his mother and his sister-in-law.

It was a lucky thing for the " Owl " that he had his ice-hut built in good time, as the sky was now menacingly dark, and suddenly we had one of those snowstorms so well known to the inhabitants at the Magnetic Pole. The driving snow was so thick that you could not see your hand before you ; inside the tents it was pitch-dark, and the drifting snow swept right into them. The " Owl " was the only one who could sit in his new hut and smile serenely ; the weather did not trouble him in the least. The snow rapidly drifted into huge mounds. The Eskimo crept out from their tents to make their preparations for the night, which, of course, is the most trying time. Had it not been so late in the afternoon, they would have built snow-huts, but as it was, they had to content themselves with erecting a wall in front of the entrance to the tent, to protect it against the fury of the snowstorm. The erection of these walls is a work of art and requires much experience, as the drifting snow, swept along by a furious storm, follows certain laws which it is very difficult to gauge accurately. I was never able to solve the problem of building one of these walls that would keep out the snow effectually ; but the Eskimo can do it. By means of a few snow-blocks he very soon diverts the whirling clouds of drifting snow, and converts the interior of the tent into a snug shelter. One after another, these sheltering walls arose, and soon every tent was protected. Luckily the storm kept steadily in the same quarter, and they were saved the trouble of having to go out again to erect new snow walls : one excursion into the open sufficed.

END OF VOL. I.

LONDON:

HARRISON AND SONS, ST. MARTIN'S LANE,

PRINTERS IN ORDINARY TO HIS MAJESTY.

ROUTE OF THE GJÖA THROUGH THE NORTH WEST PASSAGE

Made in the USA